OBJECTIVITY

To my son Thomas!

Objectivity

Recovering Determinate Reality in Philosophy, Science, and Everyday Life

TIBOR R. MACHAN

Chapman University and Hoover Institution

ASHGATE

Published by
Ashgate Publishing Limited
Gower House
Croft Road
Aldershot
Hants GU11 3HR
England

Ashgate Publishing Company
Suite 420
101 Cherry Street
Burlington, VT 05401-4405
USA

Ashgate website: http://www.ashgate.com

British Library Cataloguing in Publication Data
Machan, Tibor R.
Objectivity : recovering determinate reality in philosophy, science, and everyday life
1.Objectivity
I.Title
121.4

Library of Congress Cataloging-in-Publication Data
Machan, Tibor R.
Objectivity : recovering determinate reality in philosophy, science, and everyday life / Tibor R. Machan.
p. cm.
ISBN 0-7546-4042-6 (alk. paper)
1. Objectivity. 2. Science–Philosophy. I. Title.

BD220.M33 2003
121'.4–dc22

2003063709

ISBN 0 7546 4042 6

Typeset by IML Typographers, Birkenhead, Merseyside, and
printed in Great Britain by TJ International Ltd, Padstow, Cornwall.

Contents

Acknowledgments

I wish to thank the publisher of *Theory and Decision* (1974) for permission to use material from my paper 'Kuhn's Impossibility Proof and the Moral Element in Scientific Explanations'; Dordrecht Publishers for permission to use material from my essay 'On the Possibility of Objectivity and Moral Determinants in Scientific Change,' from Karen D. Knorr et al. (eds), *Determinants and Controls of Scientific Development* (1976); the publisher of *Transactions of the C. S. Peirce Society* (1980) for permission to use materials from my paper 'C. S. Peirce and Absolute Truth'; the publisher of *Metaphilosophy* (1993) for permission to use material from my paper 'Some Reflections on Richard Rorty's Philosophy'; the publisher of *Philosophia* (1997) for permission to use material from my paper 'Why it Appears that Objective Ethical Claims are Subjective'; the publisher of *The American Journal of Jurisprudence* (1995) for permission to use material from my paper 'Posner's Rortyite (Pragmatic) Jurisprudence,' and the publisher of *International Journal of World Peace* (1987) for permission to use material from my paper 'Terrorism and Objective Moral Principles.' All the materials taken from these works have been thoroughly reworked.

My thanks go out to Gregory R. Johnson and Jason Raibley, as well as an anonymous reader for considerable editorial help with this book. I also want to thank John Raisian of the Hoover Institution, Jim Doti of Chapman University and Dave Threshie of Freedom Communications, Inc., for their support during my work on this project.

Preface

The question of objectivity is whether human beings are capable of knowing reality just as it is. Or is there some necessary distortion in our grasp of the nature of things, a distortion imposed by the very nature of our cognitive mechanism, or by such factors as language, culture, personal ambitions, psychological disorders, and class interests? Could it be that we do not see the world at all, since we see from a particular point of view? A somewhat less drastic concern is whether we can ever satisfactorily establish that our understanding of reality is accurate, or must that always remain in doubt?

The purpose of this book is to defend objectivity in philosophy, science, and everyday life from some of its many critics. Why does objectivity stand in need of a defense? Because objectivity is a difficult ideal to serve, especially in an era of multiculturalism, deconstructionism, feminism, and diversity. People from different cultures report having radically different experiences, indeed radically different worlds. They usually claim that their experiences are as true as anyone else's. Deconstructionists tell us that we know nothing determinate beyond language, that is, that we don't know what we are talking about. Feminists often maintain that women see the world in significantly different ways from men. The idea of diversity gains much of its plausibility from the idea that people from diverse backgrounds all have their own valid ways of seeing the world. The most prominent movements in Anglo-American and Continental philosophy are against objectivity. Such figures as Richard Rorty and Jacques Derrida unambiguously deny that human beings are capable of knowing the world as it is.

Some people don't care about such philosophical questions, and that is surely their right. But they ought to care. Our nature as rational animals is perfected by caring about such things. This is why I care about them, and why I hope to convince at least a few others of the possibility of objectivity.

There is much at stake here. It matters for our confidence in how we think and live. I am not talking about reckless arrogance that is the mark of unthinking dogmatism, but of ordinary confidence that one's mind can know reality, that it is competent to face the challenges of life, that even if it makes mistakes, it is capable of accepting this and correcting them.

If we lack such confidence, then we can become vulnerable to charlatanism, quackery, and con games. Dishonest people take advantage of any lack of self-confidence, usually by claiming to possess untestable powers in various areas such as politics, psychology, religion, and even science. If we cannot know how things really are, we cannot confidently distinguish between the quacks and *bona fide* authorities (who have actually learned a thing or two about the world as it really is).

Furthermore, the quality and length of our lives depend on knowing what there is. If we can't know the world, then how can we expect to live in it? If there is no objective reality to serve as a standard, how are we to adjudicate the clash of opinions without resorting to force and fraud? The melancholy march of human history teaches this lesson: those who know well, live well. Those who know badly, live badly. The key to knowing and living well is objectivity.

Introduction

Why Will Nothing But Objective Truth Do?

> Hardly anyone will deny that it requires disciplines of mind and character to 'see things as they are' as opposed to how they appear to be and, especially in the case of psychological phenomena, how for many and subtle reasons we fantasize them to be. Nor are many people likely to deny that we must often distance ourselves from our subject so that our fears, fantasies and affections do not interfere with our sense of what is objectively the case.
>
> Raimond Gaita, *A Common Humanity: Thinking about Love and Truth and Justice* (Routledge, 1998), p. 248

> Truth, says the cultural relativist, is culture-bound. But if it were, then he, within his own culture, ought to see his own culture-bound truth as absolute. He cannot proclaim cultural relativism without rising above it, and he cannot rise above it without giving it up.
>
> Willard Van Orman Quine, 'On empirically equivalent systems of the world,' *Erkenntnis*, 9 (1975): 327–8

The Imperative to be Objective

It is a vital task of philosophy to defend objectivity: to show that human beings are able to know the world as it is and that suggestions to the contrary are wrong. If philosophers succeed in defending objectivity, not only will they show that we can know the world, but also that we know this fact as well. We will know that we are able to know.

This is not meant to suggest that all philosophers set out to show these things. Indeed, Gaita seems a bit too optimistic in his assessment of what most folks will not deny. Clearly many have the opposite intention. But these skeptics do not prove that philosophy aims at anything other than truth, any more than the existence of quacks and charlatans proves that the aim of medicine is anything other than enhancing our health.

Not all individuals have the chance to reflect on whether their minds are suited to know reality. We all know this, of course, but when we are challenged to prove it, most of us are unprepared to answer. We have what Socrates called opinions on these matters but we lack knowledge, because we are too busy living in the real world to worry about whether we know it or not. It is tacitly known or assumed, that's all.

Philosophy arises when at least some of us ask and try to answer such questions. Once philosophy emerges, all sorts of positions on objectivity are generated, aired, and considered. Most skeptical positions can, however, be easily eliminated because of self-referential problems. It is patently contradictory to claim that one knows that no knowledge is possible. If nothing can be known, then how does someone who advances that position know it to be so? Consider a prominent contemporary philosopher's opinion on these matters. Simon Blackburn, writing on Richard Rorty, speaks for many when he says,

> we have become uncomfortably aware of a large distance between truths at which we aim, whether in science or history or law or economics, or any field of interpretive endeavor, and the forces that shape our minds … Hence there is no such thing as the given, or the unvarnished truth. There are only what the Harvard philosopher Nelson Goodman called 'versions,' and the versions current at any place or time are the results of these hidden forces.[1]

The self-assurance with which such utterances are offered, in the face of the plain fact that their own claims cannot have any force if they are actually justified, is scandalous. Why should anyone actually take Blackburn to heart, given the meaning of what he tells us, namely, that everything is just a version of the truth, making his no better, no more deserving of serious regard, than any other version, including versions he would find utterly unworthy of serious regard?

In any case, how do we proceed with the task of ascertaining that objectivity can be attained by us? The discipline of philosophy is not carried out by means of controlled experimentation. People's lives, including how they are able to deal with reality, cannot be held still, under control, for experimental purposes. Any such attempt would distort any possible findings, since human beings will change their behavior in reaction to the attempt to control them. So philosophy is pursued by means of conversations and thought experiments, what is sometimes called dialectic.

At times one hears complaints that philosophy makes no progress, but this is misconceived, for it models philosophy on the special sciences and technology. Yet philosophy isn't comparable. One does not need to reinvent the wheel to use it. But in a sense one does have to reinvent philosophy. If one simply accepts philosophical principles ready made, then one cannot really be said to know them. To know a philosophical principle, one has to know the reasons for it, and to know the reasons, one has to go through the arguments oneself. Thus, in effect, each person has to reinvent philosophy for himself. Members of every generation must revisit the perennial questions of philosophy, however successful or unsuccessful the previous visits may have been. When it comes to philosophy, we are all adolescents: we will not just accept what adults tell us on authority, however true it may be.

The only progress that exists in philosophy is marginal. Issues and concepts can be more sharply defined. Novel distinctions can be introduced. Spurious distinctions can be eliminated. But no topic can be just laid aside as completely settled and immune to criticism. Precursors exist to virtually every ongoing, contemporary debate. Take the question of Artificial Intelligence, whether machines might be able to think. This is not a new concern at all. In nearly every age some suggest as much, others holding forth against the idea.

In philosophy the questions are all very basic, so much so that it is unimaginable that other fields of knowledge can proceed without some answer to them. What is truth versus falsehood? Opinion versus knowledge? Subjectivity versus objectivity? Whatever the difficulties may be in answering these questions, philosophy cannot settle for anything less than the truth.

This brief introduction should set the stage for exploring all the different forms of failed objectivity and complacent non-objectivity I will be discussing in the chapters that follow. Let me simply mention some of the most important branches of philosophy and their leading questions. I will also lay my cards on the table, stating the views that I regard as true and indicating where defenses of these views can be found. I hope thus to enable the reader to get a fairly clear, if incomplete grasp of the more positive ideas that lie behind my thinking as I criticize views which attack objectivity.

1. Metaphysics

'Metaphysics' literally means 'above' or 'beyond' physics, and 'physics' refers here to the study of nature as a whole, not merely to physics in the modern sense. The metaphysical, then, can easily be defined as the 'supernatural,' which explains why metaphysics so often strays into mysticism. But there is nothing necessarily mystical about metaphysics. Metaphysics can go 'beyond' physics – beyond the natural sciences – simply in investigating aspects of reality which cut across all the different domains investigated by the natural sciences and which cannot, therefore, be adequately grasped by the methods of one science or another.

To take a page from Aristotle: biology studies being insofar as it is alive; mathematics studies being insofar as it is quantifiable; modern physics studies being insofar as it is matter in motion. But metaphysics studies being in an unqualified sense, not being-as-this or being-as-that, but being-as-being. Metaphysics deals with the question: What is it to be as such? The study of being-as-being is also called *ontology*. Ontology deals with such questions as the metaphysical foundations of logic, the nature of causality, and whether there is one basic type of being (monism) or two (dualism) or more (pluralism).

Metaphysics has been divided into 'general' and 'special' metaphysics. General metaphysics is ontology. It is general in dealing with being as such.

Special metaphysics deals with specific kinds of being which are not adequately understood by the methods of the natural sciences. There are three traditional branches of special metaphysics.

Rational theology deals with the question of the existence and nature of God. Are there rational grounds for belief in a god? If there is a god, then what kind of knowledge can we have of him?

Rational cosmology deals with the nature of the universe as a whole. Does the universe exist in time, with a beginning and an end? Or does the universe exist outside of time, eternally? If the universe does exist in time, was it created by a god? Is order a product of design, is it basic and inexplicable, or does it evolve without design out of chaos? Other cosmological questions include the nature of space and time.

Rational psychology deals with the nature of the soul. This includes three basic questions. First, what is the relationship of the soul and the body? Are they one substance or two? Second, can the soul exist independent of the body? Can it exist before or after it comes into relationship with the body? If souls can exist independent of bodies, do they actually? Do they reincarnate? Is the soul immortal? Third, is the will free? Are all human actions determined by external causes, such that in any given choice, we could not have done otherwise? Or do human beings determine their own actions, such that in any given choice, they could have done otherwise?

In my view, ontology consists of two principles: the principle of identity and the principle of causality.[2]

The principle of identity is simply this: whatever is, is what it is. Its corollary, the principle of non-contradiction, is that no thing can have contradictory properties at the same time and in the same respect. A rose is a rose, and a living rose cannot be a dead rose at the same time and in the same respect. A living rose can die, but lives at one time and dies at another. A rose can have both living and dead leaves, but it is living and dead in different respects. The principle of identity is true of all beings, past, present, and future, actual and possible. Whatever they are, we can know in advance that they are what they are and cannot display contradictory properties at the same time and in the same respect. The principle of identity is not, moreover, merely a fact of reality; it is the foundational principle of logic. To recognize the principle of identity – that all things are what they are – is essential to objectivity.

The principle of causality is simply this: the causal properties of all beings are part of their identities. What a thing does is part of what it is. What a thing can do is part of what it is. How a thing reacts to other things is part of what it is. Because there appears to be a plurality of different kinds of beings in the universe, there is a corresponding plurality of different kinds of causality.

As for questions about the 'stuff' of the universe, I see no way of answering such questions by philosophical methods. All that philosophy can say is this: whatever the universe is, it is what it is. However many kinds of being it contains, it contains no more and no less. This is not, however, very informative. All specific information about how many kinds of beings there are and their particular natures must be determined by special investigation. The same is true for cosmological questions about space, time, and the cosmos. It is even true of questions about the separability of the soul and the existence of God. Although there are no sound arguments for these claims, we cannot rule out the possibility that experiential evidence might emerge.

The only other metaphysical question that interests me as a philosopher is free will.[3] To say that one's will is free does not deny the universality of causality. It denies only the universality of a mechanistic model of causality originating with some of the ancient Greek atomists (although some others, such as Epicurus – as reported in Lucretius' *De Rerum Natura* – made room for a variety of free will). But since there are as many different types of causal powers as there are different kinds of beings, there is nothing contradictory about asserting that one of man's causal powers is his capacity to act freely, to determine his own actions in such a way that no matter what he may choose, he could always have chosen otherwise.

Freedom of the will is closely connected with objectivity. Objectivity is a cognitive achievement. It is something that we must choose to pursue. If man is not free to pursue objectivity, then he has no control over whether the contents of his mind correspond to reality or not. Whatever he thinks, he could not have thought otherwise, and he cannot do better. Without freedom, we cannot claim that any idea is objectively true, including determinism. If the determinist is right, then he cannot claim to know it, that is, he cannot claim to have arrived at his belief by choosing to adopt truth over falsehood. He can only claim that he 'cannot help' but be a determinist, and he is helpless in the face of someone who claims that he cannot help but believe in free will.

2. Epistemology

'Epistemology' means the theory (logos) of knowledge (episteme). Epistemology is the second major branch of philosophy. Some think it is the first branch, but that cannot be, for before there is knowledge, there must be something that is known.

Epistemology is a vast field, which can be divided up in different ways. There are, for instance, epistemological questions associated with different levels of consciousness: perception, conception, and inference.

There are three basic theories of perception. *Realism* holds that we perceive a reality that exists independent of, but knowable by, consciousness. *Representationalism* holds that we perceive inner images which may or may not accurately depict objective reality. *Idealism* holds that 'to be is to be perceived,' that reality consists of mental states, that there is no ultimate distinction between the perceived world 'in here' and the real world 'out there.'

The main issue regarding concepts is the 'problem of universals.' Proper names like 'Spot' refer to particular beings. But to what do general terms like 'dog' refer? They obviously refer to kinds of things. But the question is: What is the metaphysical status of these kinds? *Realism* is the view that these kinds exist in the world, whether we recognize them or not. *Nominalism* is the view that there are no objective kinds, but that kinds are created by free acts of classification.

The same debate exists in logic as well, between *realists*, who hold that there are objective foundations for the laws of logic, and *nominalists*, who hold that logic is a matter of social convention.

A similar distinction also holds in the theory of truth. The *correspondence* theory of truth holds that a statement is true if it corresponds to the facts of reality. The correspondence theory of truth is associated with realism. Two anti-realist theories of truth are *coherentism*, which holds that a statement is true if it 'coheres' with other statements held by oneself or others, and *pragmatism*, which holds that a statement is true if it aids in man's quest to predict and control phenomena.

In all these debates, I side with the realists – realism broadly construed, since there are many positions held by individual realists which I would not care to defend. Realism is the only position consistent with the robust sense of objectivity I defend. Representationalism, idealism, nominalism, and anti-realist theories of truth cannot provide an adequate account of objectivity. They either fall into subjectivism or preserve the appearance of objectivity by redefining objectivity as intersubjectivity, that is, as social convention or collective subjectivism.

Other epistemological distinctions and debates include rationalism versus empiricism, certitude versus fallibility, and knowledge versus opinion.

Rationalism and *empiricism* concern the origins of knowledge. Does our knowledge derive ultimately from sense experience, as the empiricists claim? Or do we have access to another kind of knowledge (innate ideas, intellectual intuition) which allows us to bypass experience, as the rationalists hold? I think the empiricists are right that all knowledge requires some input from experience. But consider the principle of identity. We could not grasp this principle without some experience, but as soon as we grasp it, we see that it does not apply merely to the experiences we have had or will have, but to everything, whether we have experienced it or not.

Certitude and *fallibility* are enormously important concepts. A piece of knowledge is certain if it is immune to revision by further experience and we know that fact. By this definition, the principle of identity is certain, for we know that we will never encounter a square circle or any other contradiction in reality. A piece of knowledge is fallible if it can be overturned by experience. Even a piece of knowledge that cannot be overturned by experience can be treated as fallible if we have no way of knowing this fact. One of the main causes of skepticism is an unrealistic desire for certitude and an excessive fear of fallibility. If one demands knowledge to be certain, then we will have very little knowledge and fall into skepticism. Only if we are willing to endure fallibility can we come up with a theory that does justice to real, human knowledge.

Knowledge versus *opinion* is an essential distinction for understanding the importance of objectivity. There is no such thing as false knowledge. If something is known it is as true as it could be; that is, the judgment or statement in question corresponds with all the available facts and contradicts none. There are, however, false opinions. People believe all sorts of things that do not correspond with the available facts. Opinions are infected with subjectivity. They are as much expressions of our characters and tastes and wishes as they are of reality. Knowledge, however, is objective. It is based upon reality alone. Both knowledge and opinion vary from time to time and place to place. But knowledge varies according to the availability of objective evidence, whereas opinions vary according to a host of subjective factors, individual and collective. In any given time and place, there are many opinions about a matter but only one truth. Knowledge, therefore, is better than opinion, because it is more objective.

3. Ethics

Moral philosophy, or ethics, deals with the question of what makes life worth living. It does not deal with what we actually do, but what we *ought* to do. There are two basic approaches to this question. *Realists* hold that there are objective goods. Moral realism is often called 'natural law' or 'natural right' theory. *Nominalists* hold that the good depends upon human fiat. In ethics too I side with the realists. The nominalists are correct about the importance of human choice. But freedom of choice is not a good because we choose it to be, as the nominalists would have it. Freedom is good by its very nature, whether we choose to acknowledge it or not.

In particular, I side with the Aristotelians.[4] Ethics is an objective science, based on the facts of human nature. Each human being, like all other living beings, has a natural end which it strives to actualize in this life. Those actions and traits of character which promote man's self-actualization are good. Those which interfere with self-actualization are bad. Moral philosophy, therefore, is like medicine, which pursues the actualization and healthy functioning of the body. Moral philosophy promotes the actualization and healthy functioning of the 'soul' – one's mind, character, personality. Chapter 5 will not discuss particular actions and character traits, but merely explore whether an objective moral philosophy is possible.

There is a strong connection between moral philosophy and free will. To act morally, human beings must be able to make choices from among alternatives. As Kant noted, '"Ought" implies "can".' To say that one ought to do something presupposes that it is humanly possible. To say that man ought to choose good over evil presupposes that such choice is humanly possible.

If ethics is an objective science, are its principles universally valid? Yes and no. Insofar as ethics deals with how human beings should act, some moral principles will be universal, applicable to us all. But not all moral principles are universal, because not every moral question is about human beings as such. Most moral questions deal with how *this* individual should act in this unique set of circumstances. But there can be objectively correct answers even here. So one cannot equate objectivity and universality. The fact that some moral principles are not universal does not imply that they lack objectivity.

4. Political Philosophy

There are two senses of political philosophy: philosophizing about politics and being political about philosophy. I wish to deal with the former sense. (Leo Strauss and his students deal with the latter.) Much if not all of human life is lived in communities. These communities can be ordered in radically different ways. So the issue of what order they *ought* to possess, what makes a good or just community, is quite meaningful and vital. The debate about the best political order does not take place merely in philosophy books, but in voting booths and on battlefields. Political philosophy is, quite literally, a deadly serious occupation.

Political philosophy depends on moral philosophy. We cannot know which community is good until we know what is good for the individual as such. We can't know how we should act in concert before we know how we should act as individuals.

In political philosophy too the great divide is between objectivists and subjectivists, those who think that there truly is a best political order and those who think that there are only opinions on this matter. The most common form of subjectivism is cultural relativism, the claim that politics is necessarily parochial, the expression of a particular people at a particular time and place, with nothing amounting to principles or truths that are universally applicable to human community life as such. And there is something very tempting about this view, since it seems duly humble about our capacity to know the truth. It also seems to preserve freedom and diversity. But does it? Are freedom and diversity better served by sanctioning oppressive and chauvinistic cultures, or by arguing that they are universal and objective values, cultures to the contrary notwithstanding?

I have devoted a number of books and articles to political philosophy.[5] I argue for a libertarian or classical liberal political system on the grounds of Aristotelian metaphysical, epistemological, and moral realism, broadly construed.

5. Aesthetics

Aesthetics has come to mean the branch of philosophy that asks what art is and what good art is. More broadly, aesthetics can also deal with the beauty of nature. Why is aesthetics important? Because beauty is one of those things which makes life worth living, thus a discipline that makes us more finely attuned to and richly aware of the beauties around us contributes to the value of life. Aesthetics is perhaps the most problematic area in which to suggest that universal and objective answers can be given. And, frankly, it is the area of philosophy to which I have given the least thought. But there is nothing *ipso facto* implausible about an objective approach to aesthetics, and, indeed, no great aesthetic theorist from Plato to Kant on down has ever defended the idea that beauty is *merely* in the eye of the beholder.

6. Why Do We Need Philosophy?

Some of the greatest philosophers claim that philosophy itself is necessary for leading a good life and that the unexamined life is not worth living. There is a good bit of truth is this claim. Without an integrated worldview, it is hard to know good from evil, true paths to happiness from false ones, and to pursue the better. Not that having such knowledge guarantees success but it is hard to imagine any sustained, conscious approach to happiness, to living successfully as a human being, that can do without objective knowledge. A few may just be lucky, but reliance on such luck is itself an impediment to success.

One need not take this to mean that everyone, in order to have a chance of living a good human life, must become a professional, let alone an academic, philosopher. What is needed, however, is an individual commitment to seek the truth about the most important things and live in its light. We need to know what is good for man as such in order to know what is good for us. Knowledge alone is not sufficient, of course. We also need to act. But true knowledge is necessary to act rightly.

But what does this have to do with objectivity, objective reality, objective truth? Why can't we do without objective truth? Why can't we live with subjective opinions? Because life is too hard, too short, and too precious. Nearly all of us are pursuing happiness or well-being, however we may define it. Yet most people are not happy. Why? One reason is that they lack the objective truth about what makes life worth living. Because they have all sorts of false opinions, and every day they suffer the consequences of acting on these opinions. Objectivity is what is needed to avoid falsehood and error. Nothing less than an objective answer to the questions of philosophy will do, particularly if one thinks that philosophy is a matter of life and death – if one holds, as I do, that a sound philosophy must enter into all successful projects, including the project of living life as the individual that one is.

Conclusion

Sadly much work in philosophy has been motivated by the desire to get it right for all time and all places – and, most problematically, without any possibility for further improvement in light of experience, so that we can rest on our answers, no more work being necessary. But these are not what 'universal' and 'objective' must mean, only what certain philosophers take them to mean. This is a case of the perfect being the enemy of the good – the perfect being a kind of absolute knowledge often associated with Plato's vision of ideal forms (which Socrates apparently regarded beyond the grasp of reason), the good being something that is scrupulously based on the best available evidence, but open to constant revision and correction. Ironically, the best friends of subjectivism are those who insist upon standards of objectivity that bear no relation to man's quest for truth.

Objectivity is one of the highest and noblest aims of philosophy. Objectivity is better than subjectivity, because truth is better than opinion, light better than darkness, life better than death.

But objectivity has its enemies. Most of them, however, criticize objectivity because they do not see it, and prize it, the way I do. The aim of this book is to criticize the critics, to put objectivity in a more objective, and more positive, light.

This desire to get to what we wish to understand fully, completely, finally or so that no more work would be necessary has done philosophy much harm, mainly in leading to the widespread acceptance of skepticism. Since such a final truth is unavailable, even unthinkable – given how the world keeps going on and nothing guarantees changelessness – asking for it must lead to disappointment and dejection. And these have an impact on all aspects of life, all the lives lived under their influence.

But 'objective' need by no means have the meaning given to it by those who demand final answers. That is just what certain misguided philosophical aspirations have led many to take it to mean.

Objectivity must not be permitted to become undermined by the skepticism that arises from not being able to satisfy impossible standards, ideals of dependability. Can we depend on our minds to obtain the resources to know the way things are, or are we forever lost in a mire of self-delusion? Although I know the answer to this question to be in the positive, it takes a good deal of work to show why that answer is the right one. The main task of this book is to do that work.

On the practical front, in how these ideas are applied in everyday affairs, we can then conclude as follows. In ordinary terms, to achieve objectivity one needs to check one's own likes and dislikes and guard against their influence. One also needs to check for influences coming from outside, such as flattery on the psychological front, or obstruction of visibility on the perceptual.

To avoid bias one needs discipline and self-understanding. If I know that I am partial to those who are tall, blonde, or athletic, while working as a teacher, juror, or judge, I need to make doubly sure that what I think of their performance, the merit of their work, or their legal status isn't based on my liking (or disliking) them for irrelevant reasons. One can generalize this and decide if prejudice is unavoidable or whether discipline can overcome it.

Some argue, as already noted, that there is no way to overcome prejudice, bias, or the determination of one's culture or community when one thinks about anything. Indeed, they claim, everything we think is unavoidably influenced by such factors. Some even go so far as to claim that the very fact of having a human mind guarantees that the world won't be understood as it really is but only as it appears to us.

This and related positions are, however, troublesome to uphold consistently because they also indict the person who advances them, making it appear that one need not take the positions seriously since they, too, are just prejudiced and thus quite unreliable. In my view, in contrast, we are well able, but rarely fully willing, to rid ourselves of prejudices. We can turn our minds to consider things as others would and even as just a human being would, free of prejudice or bias, never mind specific background. A human being's mind need not be prejudiced or biased at all since it is just the sort of organ that can gain understanding without shaping the world at the same time. It is akin to when one grabs a cup, hammer, or baseball – just doing that need not have any influence on what is being grabbed. (On the other hand, if what one uses to grab something has on it paint or glue or some other stuff that can easily be transferred, the situation is different. Similarly, if one has many prejudices, biases, preconceptions one hasn't purged, one's judgments will reflect this and will be unreliable. But that isn't necessary by any means.)

Scientists, engineers, jurors, judges at athletic events or beauty pageants as well as philosophers do manage to understand the world, or at least parts of it, all the time, more or less successfully. Yet even to say that assumes that now and then success can be had, otherwise how would we even know that sometimes we fail? What would our failed efforts compare to?

Notes

1 Simon Blackburn, 'Richard Rorty,' *Prospect Magazine*, April 2003; http://www.prospect-magazine.co.uk/Start.asp. Not to mislead, however, Blackburn ultimately rejects such skeptical musings on grounds very similar to those I shall display throughout this book. For another source of criticism of anti-rationalist views, see Robert Nola, *Rescuing Reason* (Boston: Kluwer Academic Publishers, 2003).

2 Tibor R. Machan, 'Evidence of Necessary Existence,' *Objectivity*, 1 (Fall 1992): 31–62.

3 Tibor R. Machan, *Initiative–Human Agency and Society* (Stanford, CA: Hoover Institution Press, 2000).

4 See, for a detailed treatment, Tibor R. Machan, *Classical Individualism* (London: Routledge, 1998) and *Generosity; Virtue in Civil Society* (Washington, DC: Cato Institute, 1998). See, also, Tibor R. Machan, *A Primer on Ethics* (Normal, OK: University of Oklahoma Press, 1997).

5 For the position I have defended, see Tibor R. Machan, *Private Rights and Public Illusions* (Rutgers, NJ: Transaction Publishers, Inc., 1995); *Capitalism and Individualism* (New York: St Martin's Press, 1990); *Individuals and Their Rights* (La Salle, IL: Open Court, 1989), and *Human Rights and Human Liberties* (Chicago: Nelson-Hall, 1975).

PART I
Philosophy

Chapter 1

Reflections on Richard Rorty

> Why does he not walk early some morning into a well or over a precipice, if one happens to be in his way? Why do we observe him guarding against this, evidently because he does not think that falling in is alike good and not good? Evidently, then, he judges one thing to be better and another worse.
>
> Aristotle, *Metaphysics* 1006b (IV, iv, 40)

Not many academic philosophers manage to climb down their ivy walls and speak to the general public. One exception is Richard Rorty, Professor of Romance Languages at Stanford, formerly University Professor of Humanities at the University of Virginia and Professor of Philosophy at Princeton.

Rorty established his professional reputation by doing solid work in analytic metaphilosophy, the discipline in which the nature of philosophy itself is being investigated. Rorty became famous for his 1979 book *Philosophy and the Mirror of Nature*,[1] in which he condemned philosophy's traditional 'foundationalist' aspirations to know the absolute, universal, and true.

Rorty also condemned the analytic philosophical establishment for its parochialism, intolerance, and irrelevance to practical life. He praised such movements as pragmatism, hermeneutics, and deconstruction, and such thinkers as Dewey and Derrida, who are widely suspected of putting political ideology ahead of the search for truth. Rorty defends this attitude in an essay entitled, 'The Priority of Democracy over Philosophy.'[2]

Now Rorty has virtually ceased publishing in philosophical journals and has turned his hand to punditry, appearing regularly in the opinion pages of major dailies in America and England, as well as prominent intellectual periodicals such as *The New York Review of Books*, *The London Review of Books*, *Dissent*, and *The New Republic*.

1. Solidarity or Objectivity?

Rorty's view of the basic philosophical option is stated plainly: 'Solidarity or Objectivity?' This is the title of the first essay in the first volume of his *Philosophical Papers*.[3]

Rorty prefers solidarity. In the Introduction to the same volume, he attacks the idea that true thoughts must correspond to objective reality. He calls this idea 'representationalism.' Rorty's anti-representationalism seeks to 'represent objectivity as intersubjectivity, or as solidarity.'[4]

> On an antirepresentationalist view, it is one thing to say that a prehensile thumb, or an ability to use the word 'atom' as physicists do, is useful for coping with the environment. It is another thing to attempt to *explain* this utility by reference to representationalist notions, such as the notion that the reality referred to by 'quark' was 'determinate' before the word 'quark' came along (whereas that referred to by, for example, 'foundation grant' only jelled once the relevant social practice emerged).[5]

Rorty also attacks a related idea, which he calls 'foundationalism.' Foundationalism is the view that philosophy is the guardian of reason and objectivity.

My aim is to rescue objectivity and foundationalism from Rorty's critique. I will show that Rorty's efforts are not successful and are, moreover, self-defeating despite his denials that this is so.

A Brief on Foundations

I have elsewhere developed my own positive case for what I call a minimalist foundationalism.[6] Let me make a slight detour here to lay out its central element since it will help us to see how one can conceive of a metaphysics with minimal substantive commitments, one that should be the province of the special sciences.

In Ayn Rand's somewhat sketchy philosophy, which, nonetheless, I find convincing on the topic of foundations, a central place is occupied by axiomatic concepts: roughly, ideas that we cannot do without anywhere, any time (even if we don't identify them explicitly). They are basic because they point to a fact that is ubiquitous, omnipresent, even in attempts to challenge them.

Rand's concern with axioms has often been ridiculed. As one author, Leon Wieseltier, puts it, 'A = A. Big deal.' Yet her claims for the role and function of axioms are sweeping. If those claims are justified, her stress on the importance of axioms is not at all misplaced.

According to Rand, the formulation 'Existence exists' is a way of 'translating into the form of a proposition, and thus into the form of an axiom, the primary fact which is existence' (*Introduction to Objectivist Epistemology*). 'But,' Rand goes on, 'explicit propositions as such are not primaries; they are made of concepts. The base of man's knowledge – of all other concepts, all axioms, propositions and thought – is axiomatic concepts.' It is axiomatic concepts that serve as the first principles of Rand's philosophy. She defines an axiomatic concept as:

> the identification of a primary fact of reality, which cannot be analyzed, i.e, reduced to other facts or broken into component parts. It is implicit in all facts and

> all knowledge. It is the fundamentally given and directly perceived or experienced, which requires no proof of explanation, but on which all proofs and explanations rest.

This squares with common sense: we learn first and foremost what there is – not statements or propositions about what there is. And while a good deal of later learning – as well as human communication – occurs via propositions, the experiences we have provide us with plenty of brute facts that we access directly by means of our perceptual organs and minds.

In the Aristotelian tradition, two kinds of first principles exist. The first consists of basic statements describing the contents of a particular realm of phenomena that is the subject of scientific investigation. An example from geometry would be a basic definition, such as that of a point or a line. These first principles are particular to specific sciences.

The second kind of first principle applies to all sciences across the board. These principles do not provide the content of the science, rather, they rule its form. These are the principles of reasoning, such as the Principle of Non-Contradiction and its corollaries, or the principle that equals added to equals yield equals. These principles are axioms. These axioms are not stated as the premises of scientific demonstration. Rather, they identify the unspoken assumptions or presuppositions that rule and guide scientific demonstration itself (Aristotle, *Metaphysics*).

Rand's axiomatic concepts are axiomatic in this second sense. They do not appear as the premises of scientific demonstrations. Instead they (1) delimit the realm in which demonstration takes place and (2) provide the foundation for the rules of logical inference. That is to say, they are akin to the standards of measurement – for example, the meter or the yard – which cannot be questioned. To ask, 'How long is a meter?' makes no sense (except insofar as it is to be answered by conversion into another unit; ultimately, one simply points to the specific length one is taking as a unit for purposes of measurement). 'Why should one accept reality?' also makes no sense.

Axiomatic concepts do not refer to the specific content of our knowledge, but to the form that our knowledge must take. The chief difference between Rand's axioms and Aristotle's axioms is that Aristotle appears to speak explicitly of axioms solely in the context of scientific demonstration, which proceeds by induction. Rand extends the context of axioms by pointing out that they are presupposed in and ground inductive reasoning as well. Yet if one understands Aristotle's principles of being as axioms, then this supposed difference disappears.

What motivates the identification and conceptualization of the primary facts of existence, identity, and consciousness? The motivation is foundational: to ground human knowledge, to serve as a guardian against error and a corrective for it. As Rand puts it, 'Axiomatic concepts are the products of an epistemological need – the need of a volitional, conceptual consciousness which is capable of error and doubt ... It is only man's consciousness, a consciousness capable of conceptual errors, that needs special identification of the directly given …' (*Introduction to Objectivist Epistemology*).

Rand's axiomatic concepts – perhaps one of the most widely misrepresented elements of her thinking, competing only with her ethical egoism for this honor – delimit the form of knowledge in two general ways.

First, they lay down the boundaries of possible experience.

Whatever in fact we happen to experience, it will exist – that is, it will be something other than the conscious act that is our experience of it. Thus we can simply ignore the solipsists who claim that conscious acts create their objects *ex nihilo*. Furthermore, whatever we happen to experience, it will 'have' an identity – that is, it will be determinate. Thus we can ignore those who claim that they have experienced something that is both red and green (that is, not-red) at the same time and in the same respect. And finally, whatever we happen to experience, it will *be* an experience – that is, we will be conscious in experiencing it. Thus we can smile bemused at those psychologists and philosophers who claim that they are 'aware' of data which 'indicate' that consciousness, intentionality, and the given are myths.

Second, in addition to their role as boundaries, the axiomatic concepts serve as *guidelines*, as means by which we can conscientiously reach, delimit, and protect proper knowledge. We do this primarily by adhering to the principle of non-contradiction, the basic law of logic.

The identification of existence, identity, and consciousness is not 'informative,' if informative speech is arbitrarily restricted to the production of so-called 'synthetic' propositions. Nor is it informative in the sense of communicating something surprising and novel. Everyone knows these facts, or can know them upon a moment's reflection. As soon as they are stated, it seems less as if they have imparted new information than reminded us of something that we have known all along.

The purpose of identifying them explicitly is, again, not so much to inform as to underscore the basic framework of human cognition. This underscoring is reflected in the formal, intrinsically repetitive axioms drawn from these concepts.

To dismiss these axioms with the claim that they are repetitive, banal, or obvious is simply to miss the point. Of course they are obvious. If they weren't, they wouldn't be identifying basic facts.

Rand takes philosophy with the utmost seriousness. Its purpose is not to titillate with novelty but to contemplate and appreciate eternal verity, and the necessary implications of that verity. Given this conception of philosophy, repetition is not an impoverished mode of speech, but the highest and deepest.

Repetition, moreover, is especially necessary in the current intellectual climate. Rand, like Orwell, thinks that our culture has sunk to such a level of skeptical decadence that it is necessary to identify, repeat, and defend the obvious: A is A.

2. In Defense of Objectivity

First, Rorty's assertion that 'the word "quark" ... is useful for coping with the environment' has the same problem as the assertion that 'quark' *represents* reality. For it appears that Rorty has done nothing more than to switch vocabularies – unless, of course, he presupposes the position he disavows, namely, that his choice of words describes reality better than the language of representationalism. Rorty assumes that his words better ' "correspond to" or "represent" the environment' – for example, the relationship between words and things – than other people's words.

Second, Rorty's deprecation of the claim that 'the reality referred to by "quark" was "determinate" before the word "quark" came along' is self-defeating. For instance, whenever scholars discuss Rorty's views, I imagine that he feels gratified when those views are accurately represented and irritated when they are not. For example, he clearly resents being pigeonholed as a neo-Marxist or a relativist or a deconstructionist.

This can only make sense because Rorty's ideas exist in a determinate fashion prior to their interpretation by various scholars. Those scholars could, of course, simply concoct any ideas and ascribe them to Rorty, on the grounds that Rorty's views have no determinate identity prior to their interpretation.[7] But, in fact, this would be playing coy and not being charitable, as scholars are supposed to be. Nevertheless, the point is telling – on *his* terms Rorty has no basis for objecting to misrepresentation.

Third, Rorty is wrong to think that all defenders of objectivity believe that we can 'climb out of our minds' or find 'a skyhook – something which might lift us out of our beliefs'[8] so that we can determine whether or not they correctly match reality. That simply does not capture the position of many who do hold that when we think and talk carefully, we are thinking and talking about objective reality – a reality that is not being shaped or created or distorted or otherwise influenced by our minds. Thinking, of course, is not independent of our minds. We think *with* our minds, just as we perceive with our sense organs and communicate through speech and writing. But that should not undermine our conviction that we can know a determinate, objective reality.

Accordingly, when we think and talk about 'Rorty's position' on the minimum wage there is an objective reality, namely, Rorty's position, distinct from our thinking and talking. If, however, we were to talk about Hamlet's position on the minimum wage we run into immediate problems, because there is no such position.

In order to appreciate these points, it is entirely unnecessary to make use of notions such as 'getting out of our minds' to check if our minds represent reality. A less misleading metaphor is suggested by the German 'Begriff' and the English terms 'grasp' and 'grab.' When talking about cognition, 'begreifen' means to apprehend or take cognizance of, as do 'grasp' and 'grab.' Just as grasping an object in one's hand is different from manufacturing an object with one's hand, so too with grasping an object with one's mind. Just as grasping an object in one's hand does not necessarily alter it, so too with grasping an object with one's mind.[9] Just as I may pick up a coffee cup without changing it into something else, I can know a coffee cup without changing it into something else.

This cognitive sense of 'grasp' also accommodates the pragmatist emphasis on the open-endedness of the pursuit of truth. When one grasps an object, one need not have full, complete, and timeless control of it; one only needs it well enough in hand to proceed with one's purposes. Thus a cognitive grasp need not satisfy a Cartesian model of certainty, where one does not know unless one knows 'beyond a shadow of doubt,' as opposed to 'beyond a reasonable doubt.'

Grasping with the hands and grasping with the mind are subject to various kinds of failure. Just as grasping with the hands may involve greater or lesser delicacy, so too with knowing. Just as some things may slip through our physical fingers, some might slip through our cognitive fingers. Just as grasping with the hands may crush or mangle or leave fingerprints on the object, so too with knowing. But to improve our physical grasp, we do not have to 'go outside of our hands'; we simply have to take a new, more subtle and appropriate hold on things. So too with knowing.

Sometimes these failures are complete and cannot be corrected. Some point to Heisenberg's 'uncertainty principle' as proof that objectivity is impossible. If our attempt to know something changes it into something else, then it cannot be known. But even this failure is an objective fact, and it is an informative one. It teaches us that some things are too delicate and susceptible to influence to be known as they are in themselves. The same is true of physical attempts to grasp things. Smoke rings and spiders' webs are too delicate to be grasped by the hand without changing them. But Heisenberg's observations do not constitute a blanket critique of objectivity. Even if we cannot know some extraordinary things as they are, this does not imply that we cannot know ordinary things as they are. Otherwise, one could just as well argue that we cannot pick up a coffee cup because we cannot pick up a smoke ring.

Fourth, Rorty's view seems to presuppose a questionable assumption about the causal interactions of mind and world. This questionable assumption has both naive and sophisticated forms.

An example of the naive form: when I was about 12 I was sitting in my seat on a train waiting to depart from the main railroad station in Budapest. As I was looking at the people milling around outside my window, it occurred to me that these people now had a new attribute or property, something that would have to be recorded in a very detailed biography about them: to wit, I, Tibor Machan, had seen them.

But then I thought, wait a minute. This may be something that my biography should record. But surely those people will be entirely unchanged by my having looked at them.

It may well be a kind of childish egocentricity that leads one to think that knowing something changes it, so that one does not know the thing in itself, but only the thing as one knows it.

This view leads to an infinite regress, because once one accepts this position it follows, by substitution, that not only is the thing changed by being known, it will keep changing every time some additional reflection is entertained concerning it. So that not only were the people I saw changed by my seeing them, they are also changed by my reflections on seeing them, *ad infinitum.*

The more sophisticated form of this assumption arose when Western philosophy took its turn from a metaphysical to an epistemological emphasis, roughly with Descartes. It became attractive to offer mechanistic explanations for everything. Once the functioning of the human mind was to be explained within this framework, it became tempting to ascribe certain powers of efficient causation to it. Even Kant, who allowed himself a measure of dualism via his postulation of the noumenal realm, treated the sensory organs as parts of the body and saw them as exerting a causal influence upon what they perceived, thereby affecting our knowledge of the world. Accordingly, Rorty understands mind as a force that transforms the world.

But now that the universality of the mechanistic worldview is being questioned, we can also question the idea that the mind causally influences its objects. Instead, we could reconsider the earlier approach, found in Aristotle and Aquinas, that the human mind is intentionally, not causally, related to its objects. It is the object that causally impresses itself upon one's senses and then becomes the object of human awareness by the mind, not vice versa, provided the agent focuses on it.[10]

Fifth, although Rorty tries to replace objectivity with solidarity, he still cannot escape the traditional philosophical concern with objectivity. After all, one can ask: Is 'solidarity' something objective and determinate in the world? Or does Rorty have *carte blanche* to dictate what 'solidarity' means?

3. Solidarity and Individuality

If we take Rorty at his word, his politics comes before his philosophy. His praise of epistemological solidarity seems designed to support his collectivist brand of liberalism. His denial of objectivity seems designed to render his politics immune from criticism. Rorty's philosophy is particularly well designed to undermine the individualistic, Lockean form of classical liberalism that appeals to objective natural rights.[11]

Individual rights are a feature of a just human community because, contra Rorty, individuals require independence and have the power to know on their own, if such be necessary (for example, when living under the rule of a totalitarian regime wherein the bulk of official 'truth' is likely to be false);

they do not always need to appeal to social consensus, because they can, when the chips are down, appeal to objective reality (even if normally this isn't something everyone needs to do). Then, also, human progress has to have come from the initiative of individuals – collections of persons do bring forth many worthwhile achievements but it is only individuals who possess the consciousness and imagination to conceive of ideas and plans that can be implemented. And often it was single individuals who were opposed at nearly every turn by the consensus of the great majority of others who broke through with advances and only later managed to change that consensus.[12]

Not only does Rorty fail to take individuality seriously enough, he does not take community seriously enough either. It is clear that each of us belongs to innumerable communities, more one day and fewer the next, moving between them frequently. This requires of each individual to judge which deserves his loyalty, which he should value more highly, which lower, and when one and when the other, although many may simply drift with their loyalties without much thought about the matter.

Some people begin to have to evaluate whether their communities are deserving of their allegiance at an early age – when, for example, they challenge their family's various opinions, or when they may decide to stand opposed to the demands placed upon them by the laws of their society, their schools, clubs, churches, and so on. This would make it very difficult to fully account for what people do when they think and judge as expressing the views they have received from 'their community' even if we confine such communities to important, functionally crucial ones to which we belong.

Indeed, at some point those of us who are more than passive human putty need to get into the driver's seat and judge some matters for ourselves. Just what degree of contribution any one individual makes probably differs between human beings, but there is little doubt that Rorty's project of trying to reduce all individuality to some community membership is misguided.

Rorty himself, for example, refuses to conform to his own community of epistemologists – he is constantly differentiating himself from deconstructionists and radical multiculturalists,[13] not to mention the common sense of the larger community that surrounds him whose membership is certainly convinced that what they know is objectively true – as when they take account of the weather, some ailment that afflicts them, or the flat tire on their car.[14] In short, Rorty's own radical challenge of ordinary beliefs in the community or communities to which he belongs belies his account of the relationship between the individual and community. His is a clear example, in short, of how individuality sometimes overrides the community.

In addition, the many diverse communities to which we belong often make entirely incompatible demands upon us. These demands may have to be assessed by at least quite a few of us, often individually, with at most a little help from our friends but by no means with the option of abdicating individual responsibility.[15]

The human individual's creative role in life, especially concerning some of his or her ideas, is so basic that the insistence on epistemological solidarity or communitarianism simply will not manage to erase it. Human beings, in short, must come to terms with some vital options as individuals, including the option of whether to join this or that community. Not just any community will do and even those we choose to belong to must be ranked in importance, a task that we cannot simply give to yet another community. There will always be questions pertaining to the suitability of the recommended community for individuals who face the option of participating in it as well as the suitability of the order of priorities in our loyalty to the several communities to which we belong. Solidarity will not help here since the question posed is about where solidarity ought to be deployed. For that purpose what is needed is some objective standard for choosing between different possible loyalties.

This is indeed the problem faced both by radical and by conservative collectivists. It is also the main reason for the development of the very sort of rights theory Rorty finds objectionable. Such a theory, when implemented in a legal system, offers some measure of objective – rather than arbitrary – protection against being bullied into accepting the judgments of others in vital matters of belief and conduct.

Because of this, Rorty's 'moral equivalency' thesis *vis-à-vis* anti-liberal regimes will fall on deaf if not outright resentful ears among the opponents of tyrannies and dictatorships. As he puts it, 'Non-metaphysicians [of whom Rorty and, by his account, all other wise persons are members] cannot say that democratic institutions reflect a moral reality and that tyrannical regimes do not reflect one, that tyrannies get something wrong that democratic societies get right.'[16]

It is the very possibility of objective political principles that Rorty rejects, of course, not just the principles of classical liberals. (It is also arguable that his rejection requires some objective basis – for example, the objective equivalence of competing political orders.) This approach, however, leaves his own left-wing politics no more than his own (and the members of his community's) prejudices rather than capturing what is right in human community life.

It is also important that there are many different strains of belief emanating from roughly the same approach to philosophy Rorty embraces. This new turn, despite its championing of consensus over metaphysics as the ground of harmony, promises very little agreement despite its constant rhetoric in favor of solidarity rather than independent objective thinking. On the Left, we have Rorty, Derrida, Lyotard, Foucault, and Paul Feyerabend. On the Right, we have a host of Humean and Burkean conservatives admonishing us to trust the group.[17]

When the wisdom of the group reaches such radically different conclusions, the individual finds himself or herself with the challenge of making up his or her mind. When one cannot reliably appeal to the opinions of others, one is required to form one's own opinions, to open one's eyes and deal directly with reality. Solidarity then must give way to objectivity.

4. Some Initial Objections Considered and Answered

Before concluding these reflections on Rorty's views, I will address some points raised by Nicholas Capaldi, with whom I have discussed the topic for several decades and whose stance is quite friendly to this major opponent of the very possibility of objective knowledge.[18]

The first point Capaldi makes is to distinguish between the ontological realist – including Richard Rorty as Capaldi understands him – who admits that there is a world independent of the human community and an ontological skeptic. Capaldi claims that the epistemological realist (objectivist) asserts that knowledge is the grasping of a structure in objects independent of us. Capaldi sympathetically explains why Rorty and he are such skeptics. The reasons he gives are as follows:

(1a) the definition of 'knowledge' is a human convention;
(1b) no one ever specifies a criterion by which we can confirm that we have successfully grasped this alleged external structure; there is as well a Wittgensteinian argument to the effect that it is impossible even to state what the objectivist case is in a non-question-begging way. You cannot refute Rorty by saying that he cannot objectively prove his case; he has challenged the intelligibility of what it means to give an 'objective' account;
(1c) to talk about 'reasonable' doubt is to talk about the convention of what 'reasonable' means.

Capaldi then continues by claiming that

> how we understand ourselves is basic; how we understand the world outside of human activity is derivative. We can talk about the world (independent of human things) because there are human conventions by reference to which such talk can be judged. This will give no comfort to objectivists.

In response to the suggestion that what Capaldi – and Rorty – claim about how we relate to the world purports to be an objectivist or realist epistemological claim, Capaldi says:

(2a) it is not a statement about the world;
(2b) it is a statement about human conventions.

But then how do we determine the correctness ('truth', if you like) of statements about human conventions? Capaldi answers thus:

What Rorty says is that such a statement is 'true' if we 'agree' with it, where agreement has to be understood in terms of actions we perform, including but not limited to linguistic actions. Action is primary; speech is derivative. Speech never fully captures an action. Know-how is more primitive than know-that.

The thesis about justification that he espouses is 'ethnocentrism', the view that justification is relative to our practices. The defense of our beliefs against challenges by other communities (Nazis, religious fundamentalists, and so on) must always be question-begging, but this does not vitiate the defense, since no other kind of defense is available.

One will be tempted to say that all this has to be judged in terms of whether it is objectively true. But Rorty is not claiming that any of it is objectively true; he has denied the intelligibility of saying that something is objectively true. All he can do is exhibit (Wittgenstein's 'show') the practice.

Rorty's ethnocentrism appears to some to be circular and conservative, making existing practices and institutions self-justifying and impervious to 'rational' criticism, an objection also brought against other epistemological behaviorists such as Wittgenstein. This is both right and wrong. There can be criticism but it cannot be radical in nature:

(3a) 'rational' criticism is itself an inherent part of the practice (still inherently conservative);

(3b) Rorty tries to escape and provide a rationale for radicalism, but it will only work if radicalism is part of the traditional practice. Some of us would deny that it is. Marx tried to make radicalism inherent in the process, and later radicals have followed him in this respect. But this 'Marxian' move attributes a movement to the process that could only be viewed from the 'outside', but there is no 'outside'. Appealing to Kuhn and scientific revolutions will not help him because on Rorty's own terms science is derivative, not fundamental. If it is not fundamental, then it cannot be a basis for explaining any human social practice. This is an argument I make against all forms of radical post-modernism and deconstruction. Epistemological skepticism does not entail radicalism; it presupposes some version of conservatism.

Capaldi agrees with Rorty in everything except (3b) which he denies – that is, he is, as he puts it, 'a conservative in the Oakeshottean sense.' He adds 'Rorty's incoherence is in claming to make radicalism intelligible. It cannot be done. It fails on grounds of coherence and not on grounds of correspondence.'

Here is my reply to these considerations. For one, it is hardly possible for one to be an ontological realist – justifiably and thus confidently affirming a reality independent of the human mind's creations via, for example, imagination, dreams, fantasies, gossip, myths, and the like – unless epistemological objectivism is sound. In other words, that there is an independent reality is only possible for us to affirm if we can know such a reality and, indeed, know that we know it. And that amounts to objective knowledge, albeit perhaps of a sort that does not satisfy the characterization Capaldi gives it – 'the grasping (Aristotle called it "abstracting") of a structure in objects independent of us.'

Once one characterizes purported objective knowledge along the lines Capaldi does, one can easily end up a skeptic since any independent structure or essence – an idea more closely associated with Plato than with Aristotle, actually – is supposed to be timeless and perfect and, thus, inaccessible to mere mortals. It is because of this unattainable idealism about human knowledge that so many philosophers have turned to skepticism. However, this way of understanding objective knowledge is hardly universal, so the alternative between it and Capaldi's and Rorty's skeptical views does not exhaust the reasonable options.

So, when Capaldi claims, on behalf of Rorty as well as himself, that 'the definition of "knowledge" is a human convention,' the reply to this is that such a characterization of the definition is wrong as well as very troublesome to maintain without holding that this (meta-philosophical) claim is itself non-conventional, objectively true. But, going a bit slowly, what are we to make of the claim anyway? For something to be conventional is optional – as when it is a convention to wear black to a funeral or start work at 9 a.m. or give the flag to a fallen soldier's spouse. None of these are required; all could be quite different. Of course, the word 'knowledge' is conventional – it is what knowledge is called in English, but in German it is called 'Wissen' and in Hungarian 'tudas.' But what is being so called is knowledge, and that is not a convention but something some of us have learned well enough over the centuries – that human beings have concerned themselves with how they related to the world cognitively. To call our understanding of this relationship a mere convention is to misuse the term 'convention.' It is akin to calling it a convention to regard, say, arsenic as poison or AIDS as a deadly disease.

Now the reason Rorty and Capaldi appear to resort to such a drastic turn is that they work with the Platonic/Cartesian/Kantian conception of what knowledge would have to be, namely, final, perfect, total, and unchangeable truth, cognition, or conceptual awareness. But that is not what knowledge has to be – all it is, as J. L. Austin has argued, is up-to-date understanding.[19]

Has anyone ever specified criteria by which to prove such knowledge? Yet, is that even required? Capaldi notes that Wittgenstein has made a good case that 'all we can do is exhibit (… "show") the practice' that Rorty endorses, not prove it to be objectively true. But this begs the question as to what can count as being objectively true – why couldn't it be the case that a practice that is properly exhibited or shown counts as objectively real, and the claim that it is so is objectively true? Why make of 'objectivity' something so weird that it cannot be deployed as we routinely deploy it, namely, to distinguish claims that are a matter of one's prejudices or biases from those that are objectively established as those by a good prosecutor or defense attorney or by a conscientious scientist or news reporter?

Capaldi repeats a claim he makes about knowledge, only now he applies it to 'reasonable,' saying, 'to talk about "reasonable" doubt is to talk about the convention of what "reasonable" means.' Everything I noted above about claiming that the concept or definition of 'knowledge' is convention applies here, as well. As if no one had a firm, well enough established clue about what differentiates a reasonable from an unreasonable doubt, demand, proposition, or belief. It is standard fare and there is nothing merely conventional about it.

Pace Capaldi, in his defense of Rorty's epistemological anti-objectivism, all of what he tells us as well as does Rorty, is purportedly objectively true 'about the world.' If not, why bother paying any heed to such ideas – who wants simply to consider an array of parochial opinions that emanate from various tribes of human beings, unless these opinions have a good chance of being objectively true so they can teach us about what's what?

No doubt Rorty and his defenders, including Capaldi, have loads of twists and turns to offer in reply to the sort of points made here and elsewhere. Indeed, Rorty has notoriously turned away all objections to his ideas based even on their supposed incoherence and internal inconsistency, with a wave of the hand, not unlike deconstructionists have done with their critics. As if they had managed to show, beyond any reasonable doubt, that these kinds of replies are but the utterings of one community among many others with equally valid utterings – for example, the babble of those caught up in the Western intellectual or the Enlightenment tradition. But then there is no point in discussing the matter in the first place, nor, especially, in listening to what Rorty & Co. have to offer so that we might learn something about what is of interest here, namely, our ways of encountering the world and which of these manages to be the best for the time being.

Objective knowledge is impossible only if one construes 'knowledge' to mean something impossible of achievement. In Kant, for example, to know any thing-in-itself is to know it from all viewpoints, including inside out, the future, the possible, and so forth. This, indeed, is impossible to temporal agents. Their knowledge is of a sort that is firm, stable, sure, certain within the context of their cognitive capacities. Thus the reader knows objectively that reading these lines involves, in part, encountering markings that are letters and make up words and sentences, that the taste in one's mouth is that of coffee this morning, not of bourbon, and so forth and so on. These are not prejudices, preconceptions, biases, and other ways that we can be misguided about what we are aware of and constitute, thus, objective and not subjective knowledge.

A last observation about Capaldi's support of Rorty's epistemological anti-objectivism: it begins with the point he makes about the charge against Rorty that Rorty's own account of his solidarism is an objective claim alleging that certain relations obtain in reality between the human mind and the rest of the world. Capaldi's response is: 'It is not a statement about the world [but] a statement about human conventions.' This is odd since if there are human conventions, then talk or thought about them is about at least some of the world, so the point stands. He goes on to say that Rorty's claim about human conventions is true if we agree with it, where agreement has to be understood in terms of actions we perform. Yet here, too, the cart is placed before the horse. Unless some objective facts are knowable or cognitively accessible to human beings, what are we to agree on? Agreement follows our knowledge of reality. Some folks discover this or that and report it, and others, who have also discovered it, agree with them.

Capaldi and Rorty, as I have noted before, operate with an impossible model of objective knowledge and then conclude that no human objective knowledge is possible. They also embrace what Ayn Rand called the prior certainty of consciousness thesis – namely, that the mind exists logically prior to reality. This is idealism. That is what it amounts to in saying that theories of mind are about human conventions, not about the world. But the human mind is a part of the world.

5. Conclusion

Richard Rorty is the ultimate American Eastern liberal establishment intellectual, the ultimate insider. As his star has steadily risen, the quality of his arguments has steadily fallen, from dry but rigorous works in analytic metaphilosophy to little more than partisan left-wing political caucusing. He gets away with this blatant lack of objectivity because of his solidarity with other members of the establishment, who share his political aims. Why should he be concerned with the issue of objective knowledge, when his views are afforded credibility because he says nothing new, but merely confirms his admittedly prominent audience, mostly his friends, in their prejudices?

An outsider, however – someone who would challenge the consensus – lacks the luxury of being listened to regardless of what he says. Thus he must appeal to objective reasons beyond the consensus. Such an appeal is in vain, however, unless people's minds are open to objective arguments. Part of Rorty's baleful influence is that such minds are in shorter and shorter supply. But, to paraphrase Edmund Burke, one who is *incapable* of changing one's mind is in danger of losing it.

Notes

1 Richard Rorty, *Philosophy and the Mirror of Nature* (Princeton, NJ: Princeton University Press, 1979).
2 Richard M. Rorty, 'The Priority of Democracy over Philosophy' in *Objectivity, Relativism, and Truth* (Cambridge, UK: Cambridge University Press, 1991).
3 Richard M. Rorty, *Philosophical Papers* (Cambridge, UK: Cambridge University Press, 1991).
4 Ibid., p. 13.

5 Ibid., p. 5. In 'Phony Science Wars,' *The Atlantic*, November 1999: 122, Rorty has argued, 'All our controversial ways of talking are, to be sure, choices that society has made about how to classify things.' This is terribly obscure – which society; when; how does 'society' make choices? If the classifications are, as Rorty would have it, made by society, can one ask if society has made a mistake? Would society have to answer this question or might some members do so?

6 Tibor R. Machan, *Individuals and Their Rights* (La Salle, IL: Open Court, 1989).

7 It is irrelevant for my purposes here that Rorty's position is also a human artifact, as is a foundation grant. Indeed, I find the distinction between a human artifact and a natural object irrelevant to the basic epistemological concern of whether we can know reality as it is. Once an artifact is produced or created, it takes its place among the furniture of the universe, and the same issues arise concerning our knowledge of these as arise about our knowing any non-artifact.

8 Rorty, *Philosophical Papers*, 7.

9 I wish to thank Randy Dipert for pointing out that 'grasp' means 'reach for,' while 'grab' means the achievement of that reach. So, 'grab' is more akin to 'know' *vis-à-vis* the objects of knowledge. Yet, since human understanding is *sui generis*, none of these is fully adequate.

10 This is a complicated story, of course, not to be told briefly. For more, see David Kelly, *The Evidence of the Senses* (Baton Rouge, LA: Louisiana State University Press, 1986). See, also, Edward Pols, *Radical Realism: Direct Knowing in Science and Philosophy* (Ithaca, NY: Cornell University Press, 1992).

11 As an example, consider how readily he disparages the notion of natural human rights (*Objectivity, Relativism, and Truth*, p. 31). His venturing into epistemological solidarism has, to Rorty's mind, clear payoffs against any such individualist doctrine as natural rights, whereby individuals are supposed to have secured for them, as a matter of moral and political imperative, a sphere of personal jurisdiction that no community or solidarity may invade! One might indeed suspect the entire project to be but a clever strategy for capturing the political high ground for the Left. The problem is that the Right thought of it a long time before the Left did. (Consider the disdain in which most students of Leo Strauss have held rights theorists such as John Locke, and how on the matter of basic natural rights Burke and others did not much sympathize with the classical liberal tradition, even if they did favor some of the results associated with the development of natural rights theories, such as the American Revolution.) See Leo Strauss, *Natural Right and History* (Chicago: University of Chicago Press, 1953), in which Strauss contrasts the sound theory of *natural right* of the ancients with the flawed theory of *natural rights* of the moderns.

12 Later in this work I will be discussing some of the ideas of Thomas S. Kuhn, but it bears mentioning that if there is one thing his *The Structure of Scientific Revolutions* (Chicago: University of Chicago Press, 1961) demonstrated it is that many great scientific achievements come from the minds of headstrong and lone individuals. We should also recall here the well-known story of Galileo Galilei, although it is not one that has gone uncontested. See Vickie Cobb, *Truth on Trial: The Story of Galileo Galilei* (New York: Coward, McCann & Geoghegan, 1979).

13 See, for example, Richard Rorty, 'Intellectuals in Politics,' *Dissent*, Fall 1991: 483–90.

14 For examples of philosophers who see matters along these lines, see Thomas Reid, *Essays on the active powers of man* (Edinburgh, UK: Printed for J. Bell [etc.] 1788) and, more recently, Edward Pols, *Radical Realism* (Ithaca, NY: Cornell University Press, 1992) and Roger Trigg, *Reason and Commitment* (Cambridge, UK: Cambridge University Press, 1973). Among those closer to Rorty, see Thomas S. Kuhn, whose ideas I will be discussing next in this book.

15 For how this process is carried out, as understood within an individualist framework, see Tibor R. Machan, *Classical Individualism* (London, UK: Routledge, 1998), especially Chapter 14, 'Individualism versus Its Critics.'

16 Richard Rorty, 'The Seer of Prague,' *The New Republic*, 1 July 1991: 35–40.

17 Edmund Burke tells us, for example, that 'Men have no right to risk the very existence of their nation and their civilization upon experiments in morals and politics; for each man's private capital of intelligence is petty; it is only when a man draws upon the bank and capital of the ages, the wisdom of our ancestors, that he can act wisely' (quoted in Kenneth M. Dolbeare, *Directions of American Political Thought*, New York: John Wiley & Sons, 1969, p. 11). See, also, his observation that 'We are afraid to put men to live and trade each on his own private stock of reason, because we suspect that this stock in each man is small, and that the individuals would do better to avail themselves of the general bank of nations and of ages.' *Reflections on the Revolution in France* (Indianapolis: Hackett Publishing Co., 1987), p. 76.

18 Capaldi actually favors Oakeshott's conventionalist position but, as we shall see, defends Rorty, excluding only his politics. (Rorty propounds various more or less radical left-wing political and public policy ideas, certainly disparages classical liberal or libertarian notions such as human – not to mention natural – rights.) For his fully developed views, see Nicholas Capaldi, *The Enlightenment Project in the Analytic Conversation* (Dordrecht, NL; Boston, MA: Kluwer, 1998).

The following passages are taken from personal (e-mail) correspondence between Capaldi and myself, in March 2003.

19 J. L. Austin, 'Other Minds' in *Philosophical Papers* (Oxford, UK: Clarendon Press, 1961).

Chapter 2

C. S. Peirce and Absolute Truth

Peirce's Anti-Objectivism?

The father of pragmatism is often invoked as an opponent of the ideas of objective, absolute truth and an independent reality. Peirce does, however, seek to preserve mankind's experience of searching for objective truths about an independent reality by redescribing them. He redescribes objectivity as intersubjectivity. He redescribes an independent reality as what is agreed to exist by an ideal consensus of scientific investigators. I wish to argue, however, that Peirce's pragmatism does not 'work' without the concept of objective, absolute truth. Peirce is, in fact, on the side of the angels, whether he would like this or not.

Before embarking on my main purpose in this chapter, let me make clear that I am not presenting a comprehensive account of Peirce's philosophy, not even of his epistemology. It needs to be noted, also, that a good understanding of Peirce's work may well require placing it into a historical context. Peirce was doing battle with an influential school of thought of his day, namely, the Common Sense school. This outlook is most commonly associated with the Scottish philosopher Thomas Reid and had been championed in Peirce's day by Noah Porter and James McCosh.[1] Peirce was critical of this school and yet, as I attempt to argue here, there are elements of his thought that cannot shake the realism that Reid had championed.[2] (Of course, in that respect my own inclination is toward Reid's view, rather than that with which Peirce is closely associated, namely, pragmatism – or pragmaticism.) But such realism that I attribute to Peirce below has no affinity with the theological aspects of the Common Sense school.[3]

So, let me now turn to the substance of my concern with Peirce's views. What is an 'objective, absolute truth'? It is a proposition that identifies facts of reality that are basic, universal, and inescapable.

Such a truth is objective because it identifies something that exists and is what it is independent of the proposition. If one states that the sun is shining today, and it is shining today, the sun would be shining no matter who made the proposition or whether the proposition had been uttered at all. This view is known as the 'correspondence' theory of truth.

A truth is absolute if it states a fundamental, universal, and inescapable fact, a fact that holds no matter what other facts might also exist. Not every objective truth is absolute. The sun, after a time, will no longer shine. But all absolute truths are objective, else they would not be truths.

Peirce's Pragmatism

My concern here is with Peirce's fallibilism, understood as a version of epistemological probabilism – that is, the view that however well some belief is established, it could still use additional support and is only *probably* true. Peirce explained that 'we can never hope to attain by reasoning … [to] … absolute certainty, absolute exactitude, absolute universality.'[4] Whereas the correspondence theory of truth implies that truth is an all or nothing property, Peirce's probabilism makes truth a matter of degree.[5]

Peirce believed that finding some 'absolute assertion'[6] would lead us 'to set up a philosophy which barricades the road of further advances toward the truth.'[7] Peirce held that within each context of inquiry – scientific, philosophical, legal, and so on – it is possible to find basic propositions with what can be called axiomatic status *for the context*, but there are no absolutely basic axioms. That is, there are no propositions that are true beyond any reasonable doubt, propositions that cannot be improved upon by further investigation.[8]

What did Peirce consider the proper way of achieving truth *per se*? He held that 'The opinion which is fated to be ultimately agreed to by all who investigate, is what we mean by the truth, and the object represented in this opinion is the real.'[9] Truth, in short, is what investigation *aims at.* But this implies that metaphysical and epistemological 'first principles' – basic truths about reality and fundamental categories of the understanding – cannot be possible truths, for these are the *presuppositions* of investigation, not their aim.

It is important to point out that Peirce regards truth by correspondence in a Platonic fashion, that is, as requiring final, perfectly accurate, and timeless propositions to which no modification could ever occur. This conception of truth is a mistake. Truth is, rather, an attribute of propositions that correctly track the features of an ever-changing world, that capture what is the case as well as we can for the time being. Just because a proposition does not tell the final story about reality, that is no reason to treat it as merely probably true. It only appears 'probably' true when measured by the criterion of the final truth, the last word, a synoptic vision of the whole.

Human beings are, of course, fallible in the sense that any *person* making a true judgment or proposition could be mistaken. It also makes sense to describe *propositions* as fallible, insofar as they can fail to be true. Finally, it makes sense to describe the *evidence* offered in support of a proposition's truth as fallible, insofar as it can fail to establish truth. But the fact that *truth-sayers* are fallible does not imply that *truths* are fallible. The fact that *propositions* and the *evidence* offered for them are fallible does not mean that *true propositions* are fallible.

Peirce's Inconsistency

Peirce is wrong to deny that some propositions are absolutely true. Indeed, Peirce himself accepts some absolute truths. He presupposes certain metaphysical first principles without acknowledging them. These first principles are presupposed by all other philosophers as well. Indeed, they are presupposed by every human being in every cognitive act.

Peirce argues explicitly that logic is indispensable for understanding reality. And if logic presupposes certain metaphysical first principles which are absolute and objective, then Peirce himself tacitly admits the existence of absolute, objective truths.

The prime example of these metaphysical first principles is the principle of non-contradiction. This principle is true of all possible facts, and its truth is presupposed by every cognitive act, from philosophy to science to balancing one's check book. No matter what there is, it is what it is, and it cannot display contradictory characteristics at the same time and in the same respect. This applies to all past experience, all present experience, all future experience – and even to things we will never experience. If any truth has claim to both fundamentality and absoluteness, it is this. Here is a truth that is not as context-dependent, fallible, and revisable as Peirce would have it. Since the principle of non-contradiction is true, Peirce's view of the possibility of absolute truth must be mistaken.

It has been suggested that there are logics in which the principle of non-contradiction is not required at all, but this is a misunderstanding. Presumably these logics are what they are. Furthermore, they are not Aristotelian logics at the same time and in the same respect as they are non-Aristotelian logics. Furthermore, the advocates of such logics never seem to have trouble distinguishing between their logical systems and the Aristotelian approach. Because of this, they are in fact Aristotelians in spite of themselves, Aristotelians in the very act of denying it.

The Question of Logical Formalism

Now one might grant that the law of non-contradiction is presupposed in all discourse and intelligible activity, yet maintain that it is but a formal principle and thus does not constitute an absolute truth about the world. Peirce himself tells us that 'every logical principle considered as an assertion will be found to be quite empty; considered as expressing truth, it is nothing.'[10] This seems to imply that Peirce shares the modern separation of logic and its principles from the domain of substantive truth. Accordingly, Peirce presents himself as a formalist, someone who would deny the (alleged) necessary connection between logic and reality as advocated by Aristotle and others who take an ontological view of logic. Yet, in another place Peirce does not seem to accommodate the formalists so readily:

> A logical principle is said to be an *empty* or merely formal proposition, because it can add nothing to the premisses of the argument it governs, although it is relevant; so it implies no fact except such as is presupposed in all discourse ...[11]

This remark is seriously qualified, so we might wonder what Peirce thought was presupposed by all discourse. I will return to this point later in my argument. For now I will accept Peirce's earlier characterization of logical principles as the one that accords best with his fallibilism and his conception of the nature of truth.

The distinction between formal and substantive truths is a thesis which requires proof. A proof of this distinction may not presuppose the distinction that it seeks to prove. So such a theory must be proven by arguments whose premises are true, pure and simple. If Peirce is right and no absolute assertion can be found on which to base the proof of his distinction, then the proof that logic is merely formal must result in fallible conclusions. This shows that the formalist thesis, as defended in a Peircean framework, is very unstable. Thus we are entitled to consider alternative frameworks.

This argument does not show that the law of non-contradiction is an absolute substantive truth, but only that the Peircean argument that it is a formal and empty truth is ill supported. Given the kind of premises he allows, Peirce simply does not offer a theory that discourages us from searching for alternatives.

Advancing the Objectivist Case

The positive elements of this discussion flow from the negative points I have just made. By accepting the possibility of *proving* some conclusion, however fallible, Peirce must philosophically explain the nature of a proof without recourse to the substantive/formal distinction. At least he must accept that a proof is a valid way of settling intellectual questions. And when asked to explain what 'valid' means, he must accept that 'validity' is defined, in part, by reference to the concept of truth – that is, a valid argument is one in which if the premises are true, the conclusion must be true. The idea of 'truth' is presupposed, conceptually or epistemologically, in the understanding and employment of proofs, in proving something, and in rationality itself.

So logic presupposes the meaningfulness of truth, and this truth cannot as yet be split into two types, formal and substantive, since any proof of that distinction already requires logic in which truth is presupposed.

It is often suggested, of course, that merely having the concept of truth gives us no reason to think that anything is true. Peirce, however, appears to have rejected this objection. It also appears less cumbersome to think that some truths are actually available than to think that the concept of truth is meaningful, yet does not refer to anything real. Furthermore, I suspect that the theory of meaning presupposed by this thesis relies on the formal/

substantive distinction: the concept of truth is 'meaningful' in terms of its relationship to other concepts and does not require a substantive referent.

If we are not to be left with the cumbersome view that truth is presupposed in logic and rationality but unavailable to us, recall Peirce's assertion that, 'a logical principle … implies no fact except such as is presupposed in all discourse.' Here Peirce seems to abandon his formalist stance and appears to lend support to the view I am defending, namely, that the principles of logic do presuppose substantive truths. Here Peirce refers directly to facts, not merely to propositions presupposed by logical inquiry. But what fact is he referring to? The prime candidate is the principle of non-contradiction, viewed as a substantive rather than merely a formal truth. The principle of non-contradiction is true of the world and its truth is presupposed in all attempts to prove anything. It is, furthermore, not subject to any rational doubts regardless of what context of inquiry one might enter, for its truth is presupposed even in the act of arguing that it is dubious. Thus it appears to be just the kind of absolute substantive truth that Peirce thought unavailable.

Peirce also thought that absolute truths were barriers to the further pursuit of truth. Is this true of the principle of non-contradiction? Yes and no. Some epistemological guidance is provided by the fact that *nothing can contain contradictory qualities at the same time and in the same respect.* (This, despite some of the careless science reporting about the alleged implications of the wave–particle duality of quantum mechanistic particle physics.) Scientists, detectives, and ordinary people everywhere will avoid confusion if they keep this fact in mind. If one were to think that truth might violate the principle of non-contradiction, then advances toward the truth would be barred by a philosophy in which the principle is taken as an absolute assertion of fact. And to see whether such a conception of truth could be advanced we must consider whether a distinction between truth and falsity could be maintained in terms of it. That requirement, namely, that the distinction be available, is primitive, if not in every object language (for example, in some alternative logics), then at least in the pertinent metalanguage. And it is clear that such an idea of truth renders this primitive distinction unavailable except by arbitrary fiat.

Thus, indeed, advances toward truth, conceived along mystical or irrationalistic lines – if that can be called 'conceived' at all – would be closed off by a philosophy that holds the principle of non-contradiction to be an absolute. But within the framework of the principle of non-contradiction there is a whole universe of non-contradictory facts to be discovered, which is not much of a limitation by anyone's measure except the mystic's and irrationalist's.

Peirce and Objective Truth

Now we can return to Peirce's conception of truth and see how it is affected by the considerations advanced above. Peirce appears to admit that logic presupposes certain objective and absolute truths, like the principle of non-contradiction. If this is correct, then the statement of such truths has to be possible independent of the agreement of the 'community of investigators,' for such agreement could only emerge through extensive discussion, and such discussion presupposes the principle of non-contradiction.

This point is implied, also, by the epistemological dependence of logic, proofs, rationality itself, on the existence of some truth(s).[12] The axioms of metaphysics are objective truths that are epistemologically fundamental. And this would imply that some truths are not at all dependent upon agreement among investigators.

But here we arrive at a serious challenge to Peirce's philosophy and to the entire pragmatist and operationalist framework. We may have to accept that the collectivism implicit in Peirce's epistemology and social thought requires revision because it is possible for individuals to recognize the truth outside of, and even in conflict with, communities of inquirers. This does not, however, deny Peirce's insight into the invalidity of an atomistic conception of individuals. All that need be appreciated is that, despite the enormous possible value of communal truth-seeking, communities can be wrong and individuals right. Individuals who know, or at least make use of, the absolute truths presupposed by rational inquiry can both benefit from the communal truth-seeking and check the results of that effort.

Notes

1 James McCosh, *The Scottish Philosophy, Biographical, Expository, Critical, from Hutcheson to Hamilton* (New York: R. Carter, 1875).

2 However, as one reader of this manuscript noted, the American Common Sense school was strongly religious. Moreover, it needs to be noted, also, that Peirce was a prominent logician of his day and challenged various aspects of Aristotelian logic – a challenge that had triumphed during the early and middle twentieth century. Although this is not something I will take up in this work, some of that challenge and triumph may be said to have been successfully addressed more recently, for example, in Fred Sommers, *The Logic of Natural Language* (Oxford: Clarendon Press; New York: Oxford University Press, 1982) and was not unopposed even in the heyday of the new school. See H. W. B. Joseph, *An Introduction to Logic*, 2nd edn (Oxford: Clarendon Press, 1916).

3 One of those whose epistemology comes close to what I would embrace, Edward Pols, also develops certain ideas sympathetic to a theological perspective, although I cannot exactly identify it from what I have read in his works *Radical Realism* (Ithaca, NY: Cornell University Press, 1992), *Acts of Our Being* (Amherst, MA: University of Massachussetts Press, 1982), and *Mind Regained* (Ithaca, NY: Cornell University Press, 1998). For now it should suffice to note that my own views are closer to the secular neo-Aristotelians Ayn Rand and John Searle (especially as regards the latter's theory of mind).

4 Charles Sanders Peirce, *The Philosophy of Peirce. Selected Writings*, ed. Justus Buchler (New York: Harcourt, Brace & Co., 1950), p. 56.

5 It was Karl Popper who made the fallibilist theory the prominent one it is now. See, most pertinently, Karl Popper, *The Logic of Scientific Discovery* (New York: Basic Books, 1959).

6 Peirce, *Philosophy*, p. 55.

7 Ibid., p. 54.
8 Ibid., pp. 47–59, 354–60.
9 Ibid., p. 38.
10 Ibid.
11 Ibid.
12 Actually, the existence of some fact(s), proposition(s) of which would be objectively true – that is, true by correctly stating what is the case independently of the agent's influence on or attitude toward the state of affairs being correctly stated.

PART II
Science

Chapter 3

Thomas Kuhn and the Objcctivity of Science

Introduction

There is no better measure of the impact of Thomas Kuhn's *The Structure of Scientific Revolutions* than a letter to the editor in the 31 May 1999 *Newsweek*, which begins, 'As a middle-aged dyed-in-the-wool country-music lover, I had a complete paradigm shift the moment I first heard Andrea Bocelli's voice.' Not that Kuhn was exceptionally clear about what he meant by 'paradigms' and 'paradigm shifts.' Still, most people take a paradigm to be a general framework within which particular problems in a science – or any discipline or inquiry – will be approached for a while. Interestingly, Kuhn's own position came to be a kind of paradigm for the study of scientific development.[1]

What does Kuhn imply about the nature of objectivity? He offers the most serious challenge to objectivity in the twentieth century in his *The Structure of Scientific Revolutions* (1962; 2nd edn 1970). Kuhn did not find fault with ethical, political, or aesthetic objectivism, which would not have been news in this era. Instead, he casts doubt on the objectivity of science.

Prior to Kuhn there had been a broad consensus concerning the aims of science. It was supposed to strive to obtain a clear, objective, unbiased and unprejudiced identification of some particular realm of reality, for example, the physical, chemical, musical, economic, or psychological realm. Once scientists working in a particular field arrived at interesting hypotheses, they would construct experiments or collect data to determine whether or not these corresponded with objective reality. They would then report their findings to other scientists, who would try to replicate their experiments and build upon their results. Genuine scientists look at their theories from reality's point of view, adopting or discarding theories according to their correspondence with reality. They do not look at reality from a theory's point of view, seeing only those facts that confirm their pet opinions.

Kuhn will have none of this naiveté. He argues that scientists see all of reality through the lenses of their theories, and therefore never encounter objective facts upon which they stand for an independent perspective on reality from which to evaluate their theories. From this Kuhn infers that the choices made at these stages *cannot* be justified, leaving the impression that scientists may not be responsible for acting in objectively sound, right, correct ways – as opposed to objectively wrong, unsound, incorrect ways.

Kuhn's judgment is that, in the last analysis, so-called subjective aesthetic considerations determine what we affirm to be the case.

Kuhn significantly changed people's conception of what scientists do. Educated people now accept that science is perspectival, that paradigms govern even the most technical aspects of scientific thought and research. From time to time these paradigms change when they stop being useful for purposes of organizing the way we see things in the world. The new paradigms may be quite different, though not necessarily closer to the truth or more faithful to the facts of reality. Indeed, some of what Kuhn wrote suggested that different paradigms get us to see different realities or worlds. This in turn suggests that science does not provide an objective understanding of the world.

Kuhn's characterization of science is quite problematic. To show this, I will examine the philosophical underpinning of Kuhn's ideas and his support for his conception of science.

Kuhn and Wittgenstein

Kuhn observes that scientists 'can agree in their identification of a paradigm without agreeing on, or even attempting to produce, a full interpretation or rationalization of it ...' (1970: 44). He justifies this sort of procedure in science by citing Wittgenstein:

> For Wittgenstein ... games, and chairs, and leaves are natural families, each constituted by a network of overlapping and crisscross resemblances. The existence of such a network sufficiently accounts for our success in identifying the corresponding object or activity ... (1970: 44)

Kuhn is preparing us for envisioning that 'Something of the same sort may very well hold for the various research problems and techniques that arise within a single normal – scientific tradition' (1970: 45).

These remarks also characterize Kuhn's own approach. Kuhn's 'aim is a sketch of the quite different concept of science [from that of 'a constellation of facts, theories, and methods'] that can emerge from the historical record of the research activity itself [within the history of science]' (1970: 1). So the method of Wittgenstein is evidently thought by Kuhn to be appropriate to what he himself is doing. He does not, for example, define the concept 'science,' but talks of the 'image of science.' Instead of discussing the *nature* of paradigms he simply tells us what he 'takes' these to be. Both of these cases clearly accommodate his interpretation of Wittgenstein's idea that talk of 'the nature of X' in any familiar way is inappropriate. Kuhn never considers any alternative to the idea of natures and definitions Wittgenstein criticized. This, too, indicates his full rejection of objectivism in formulating definitions.

Kuhn's problems start here. It seems unwise to extrapolate Wittgenstein's conclusions about the nature of identification of objects like chairs

(Wittgenstein, 1953, 1973; Stroud, 1966; Machan, 1973); we must take care not to misinterpret the idea that when we try to identify the essential features (components, properties, aspects, and so on) of such objects (and anything else we may want to); all we are entitled to expect is a statement of 'a network of overlapping and crisscross resemblances.' I am sure we can and ought to do better. Wittgenstein seemed himself to be clearing a path to such better solutions to the problem of definitions, criteria, standards of judgment, and so on (Stroud, 1966; Wittgenstein, 1969).

Kuhn, however, takes Wittgenstein to be offering a clear alternative to the empiricist approach with the idea of family resemblance. Kuhn talks of the implications of this idea for the problem of identifying a paradigm. He says that the Wittgensteinian idea shows that no justification for such an identification could be provided – instead we have 'interpretation' and 'rationalization' (!) (1970: 44). In the Preface to his work, Kuhn follows faithfully this conception of what a scientist needs (and does not need) so as to produce 'the particular conclusions he does arrive at …' (1970: 4). He says that these 'are probably determined by his prior experiences in other fields, by the accidents of his investigation, and by his own individual makeup.' He tells of 'that element of arbitrariness' that is 'compounded of personal and historical accident … [which] … is always a formative ingredient of the beliefs espoused by a given scientific community [and, he says elsewhere, by the individual] at a given time' (1970: 4).

Accordingly, Kuhn gives us a plethora of data about himself as he launches into his own project – sources of insight, influence, guidance, 'random exploration,' and other bits and pieces. These, by his own account, should help us to appreciate (though not justify) Kuhn's objective to urge a change in the perception and evaluation of familiar data as well as a 'new image of science' – that is, to help us witness identification without justification.

Am I fair to think that Kuhn's approach isn't just self-referential by accident but also by design? Clearly he considers himself to be doing history as well as sociology of science based in part on 'historical evidence' and resulting in something very close to a law of the history of science – for example, when he says that 'if I am right that each scientific revolution alters the historical perspective of the community that experiences it, then that change of perspective should affect the structure of postrevolutionary textbooks and research publications' (1970: ix).

I would not be surprised if Kuhn conceived the recent work in the history, sociology, and even philosophy of science as undergoing just the kind of change he is talking about in consequence of the revolution his work – and that of others suggesting his ideas in greater and greater numbers throughout the last few decades – produced. However, this would pose a problem for Kuhn when talking about being 'right,' in light of some of his epistemological ideas – as I have argued elsewhere (Machan, 1974b) and will come to later in this discussion.

The above considerations set the stage for my criticism of Kuhn's sociology (and history) of science. I will not deal with Kuhn's specific renditions of his historical data, although I believe that for him to come even close to providing a meaningful[2] justification for his conclusions about what actually went on in science throughout history, he would have to do more than *cite* widely discussed cases. This is because Kuhn would have to show that nothing else but what *he takes to have happened* (in terms of his own paradigm) did happen – scientists *actually* did what they and/or their biographers say they did, that is, whether things in fact occurred as they would have to for Kuhn's conclusions to be right.[3] Obviously I cannot make any headway here on such a project. Instead I am concerned to challenge the soundness of Kuhn's approach by pointing out its methodological and epistemological faults.

In trying to criticize Kuhn, one faces a recurring problem. By Kuhn's own account, any criticism *must* begin with arbitrary standards.[4] Although Kuhn directs this point at the dynamics of scientific thinking and research, it is evident from what I have indicated above that Kuhn's own approach falls under the kind of thinking and research scientists carry out. Before anything even mildly hopeful can be said in criticism of Kuhn, we need to see if the philosophical underpinning of his approach to the history of science is correct.

This will take us directly to Kuhn's epistemology, as well as to the metaphysical implications of his approach. For while Kuhn eschews all epistemological argumentation in his book, the underlying epistemology we find can make it clear what goes on at later stages of his discussion. Moreover, it is not as if Kuhn were oblivious to such issues. He only introduces them without sufficient detail and self-criticism. By this introduction, however brief it may be, Kuhn opens the door to cross-examination on this topic.[5]

Kuhn's Misuse of Wittgenstein on Family Resemblance

Kuhn brings in Wittgenstein's point about family resemblance very early, and quite understandably so. He aims to rest a good deal on it. But there is reason to believe that Kuhn reads too much into this Wittgensteinian point. Kuhn thinks Wittgenstein presented us with an impossibility proof about definitions of concepts.

But what Wittgenstein did, in fact, show was the error of a particular model or 'picture' for what a definition must be. He showed that taking a definition to be a fixed, immutable, unchanging standard is a mistake. He objected to conceiving of criteria in such a way that once they were correctly understood and identified, no more could possibly be learned about them. Wittgenstein realized that this view held it impossible that we *could* find it necessary to later revise our criteria for identifying what something is. He clearly seems to me to have been intent on dislodging a variety of Platonism,[6] where the model of what knowledge and truth must be is restricted so that only certain kinds of things are capable of being known.

This, Wittgenstein seemed to have realized, amounted to a prejudice – a case of prejudgment about *what* human beings can know and about the nature of knowledge itself, what *could* constitute knowledge of reality. (Incidentally, the Platonist *model* need not be restricted to idealism, so Wittgenstein's objections could hold in all cases where the model guides investigation, independent of the particular account of knowledge by which it is satisfied.)

Kuhn is not unjustified in invoking Wittgenstein in connection with his criticism of the positivist ideals of knowledge, truth, and science. Wittgenstein's argument against an attempt to find what 'all games and only games have in common,' namely certain 'features' and 'elements,' (1953: 66) is placed within the context of a certain idea many philosophers – including Wittgenstein himself at one time – had about *what* these features *must* be. He does not conclude that under any interpretation of 'common' it would be impossible to require common features of things, and so on, to be identified as one *kind* of thing, and so on. He says that 'we see a complicated network of similarities, etc.,' but this would appear to apply to efforts that aim *at perceiving common* features.[7] In Wittgenstein's context the conclusion about coming up only with family resemblances need not imply that we could *not* do better by producing a different account of the nature of criteria for identifying the essence of games. Indeed, Wittgenstein goes to some lengths to provide clues for the formulation of an alternative approach, one I think expressed most cogently by Barry Stroud (1966).

Kuhn drops the context of Wittgenstein's discussion. He then extrapolates (what he wrongly takes to be) Wittgenstein's point to the subject matter of the justification of a paradigm. So Kuhn begins on the wrong foot epistemologically. Then Kuhn appeals to several other notions to support his mistaken beginning – for example, Piaget's experiments, the difficulties associated with making correct distinctions in the cases of cards that have their colors switched, and the duck/rabbit drawing (which he introduces at one point but appears to drop at a later one) (1970: 63, 113).

This allows Kuhn to speak of the ultimate incommensurability (1970: 44) between what he maintains are equally adequate *scientific* paradigms. For Kuhn does end with an *impossibility* proof. He warns that his position does not 'suggest that new paradigms triumph ultimately through some mystical aesthetic.' Nevertheless he proceeds as follows:

> But if a paradigm is ever to triumph it must gain some first supporters, men who will develop it to the point where hardheaded arguments can be produced and multiplied. And even those arguments, when they come, are not individually decisive. Because scientists are reasonable men, one or another argument will ultimately persuade many of them. But *there is no single argument that can* or should persuade them all. (1970: 158, emphasis added)

Because there is no single argument of that sort, 'A decision between alternate ways of practicing science is called for, and in the circumstances that decision must be based less on past achievement than on future promise ... A decision of that kind *can only* be made on faith' (1970: 158, emphasis added).

Actuality, Possibility, and Impossibility

It is with this element of impossibility that I want to take issue. Kuhn does not *show* that 'there is no single argument that can or should persuade them all,' or that 'A decision of that kind can only be made on faith,' albeit he reads the historical material to be *indicating* such an impossibility.

But, first, Wittgenstein's *Philosophical Investigations* do not offer an impossibility proof. Even if the family resemblance argument works – and there are grounds to claim that it does not[8] – the context of that argument is misread by Kuhn. Second, Kuhn is talking about what *persuades* scientists, and persuasion is not what is at issue in arguments, only in seeking converts.

What is at issue is whether some (set of) arguments might be conclusive in the context of solving some problem or settling some dispute. Perhaps many instances of paradigm introduction *have not* been accompanied by conclusively sound or unsound justifications. But this does not prove that they *could not* be so accompanied. There are instances in most areas of science where even from within the existing context more care and attention could have precipitated better selections.

Jean Rostand (1960) tells that the history of N-rays involved both honest error *and* wishful thinking. He also chronicles the story of radiesthesia and metapsychics, the first generally avoided by reputable scientists but the second still going quite strong. In these and other cases, a paradigm gained a significant degree of success unjustified by the relevant considerations.

What is crucial, of course, is that the 'relevant' considerations be identified, and Kuhn does not help us understand in the least when scientists have *good* reasons for following through on paradigms, what with his consignment of decisions or choices to the area of faith. This Kuhn seems to do without researching the kind of considerations that might have to be appealed to at the crucial stages of drastic developments within some field of science. Are we to understand that these stages involve no epistemological issues? That nothing of metaphysics could have a bearing on what will be taken to be good reasons for making selections between competing paradigms? Or if we should consider those factors, does it not follow from his views that in *all* areas mentioned, and others that could be relevant, we can only make a decision on faith? Granting that in periods of normal science there is a kind of 'objectivity' to scientific reasoning – based on the accepted paradigm and what counts as good reason within it – must we accept Kuhn's broader, or what I want to call transcendental, subjectivism because at the critical stages no pre-established basis for selection is readily available, that is, since none are agreed to or identified just then?

This is no small point. Kuhn and I may both agree that scientists themselves might not *actually* bother with justifying their paradigm choices, especially when such choices accord with some widely accepted doctrine about the nature of knowledge, argumentation, evidence, or other extra-scientific and more properly philosophical beliefs. But in Kuhn's scheme these too *must* end in arbitrariness.

Kuhn is actually proposing a variety of (Peirce's) pragmatist epistemology[9] and implicitly, at least, affirms the impossibility of identifying some metaphysical fact that can function as the ground of justification. Yet he has argued none of this, leaving us, nevertheless, with a very powerful polemic. I intend to point out how crucial, paradoxical, and, indeed, dangerous the consequences of this view are.

Kuhn's Self-Referential Arguments

But for now let's consider another feature of Kuhn's method of argumentation. Kuhn makes several of his points on the basis of evidence drawn from history, psychology, and sociology. Evidence from history means case histories in Kuhn's context. But this is not so with evidence from psychology and sociology. Kuhn himself tells us that the weight, significance, and even the factual content of the claims of scientists are paradigm-determined. He tells us that in the role of 'a vehicle for scientific theory … [a paradigm] … functions by telling the scientist about the entities that nature does and does not contain and about the ways in which those entities behave' (1970: 109). Kuhn believes that scientists who work within a paradigm are limited by it as to what they may admit as evidence, as to what they are entitled to call a fact, and regarding what they may take to be an entity of the sort involved in their investigations.

Before turning to the consequences of Kuhn's ideas for his own method and results, it should be mentioned that Kuhn assures us that 'Whatever he may then see, the scientist after a revolution is still looking at the same world' (1970: 129). But this is very difficult to reconcile with the main trends in Kuhn's theory, since he also says that during revolutions scientists 'see new and different things when looking with familiar instruments in places they have looked before' (1970: 111). This leads to difficulties. By not entering into the philosophical issues involved in these remarks, Kuhn renders it impossible for us to understand him clearly. What is it to be looking at the *same* world? More specifically, how could one tell *that* if, necessarily, we all may see the world differently?

When we turn these puzzling ideas upon his own 'scientific theory,' Kuhn immediately faces difficulties. The evidence for his theory is no evidence at all. Instead it must be construed as a case of the world (being looked at) 'seen differently from the way it had been seen before.' We have to ask whether the material cited by Kuhn could ever be invoked to support him, whether they are material facts about the Priestley/Lavoisier affair, Piaget's discoveries about children's perceptual development, or problems about telling the difference between cards ordinarily colored one way when the colors are switched.

Kuhn could reply that he is merely trying to make sense of the world by his interpretation of some important events of concern to us. But this will do for data from the history of science only, not for Piaget's data (1970: vi, 112, 113), for in the latter case Kuhn is not trying to *make sense* of what Piaget discovered but is using the discovery to make sense of the history of science. What he says about seeing the world via a paradigm must apply to Piaget also.

Once a paradigm switch occurs in perceptual psychology, and Piaget's facts take on a different significance (as determined by it), not much is left of Kuhn's evidence. The same can be said about the card problem and the duck/rabbit case. These too would have to be reinterpreted if a paradigm switch obtains within epistemology and psychology. And when we push the self-referential point further, we are left with no materials that can, in principle, give support for Kuhn's own position – not even in some kind of network of overlapping and criss-cross resemblances.

Truth and Faith in Kuhn

Let us now turn to a philosophically central idea in Kuhn's system, namely the model of 'being right' or truth that he appears to invoke as he drives home his conclusion that decisions about which paradigm ought to be selected 'can only be made on faith.' This last issue is most crucial for purposes of assessing Kuhn's metascientific theory, especially its impact upon a concern with how best to conceive of science as a human activity. Kuhn here is unfortunately giving support to the absurd and damaging contention that the human mind, in this case that of any scientist, is unable to make reliable, justifiable, conclusive, and correct identification of what there is in reality; he is affirming the impotence of the human mind in the matter of learning whether we have identified reality correctly.[10]

Kuhn says, late in his book, that he rejects the view that science evolves toward anything (1970: 160, *passim*). He concludes this from a criticism of the (so-called) positivist notion that science began with ignorance and is moving toward some conclusive, completed truth about nature. His account rejects this idea in favor of 'a process of evolution *from* primitive beginnings … whose successive stages are characterized by an increasingly detailed and refined understanding of nature' (1970: 170). How he is able to say this much is unclear, but for him the question of whether this understanding is valid cannot be asked except from the viewpoint of a paradigm – one that is selected, in the end, on faith.

A word here about what Kuhn might mean by 'faith.' He does not discuss it. The idea is generally used to mean belief without evidence.[11] Existentialists have used the idea to mean an act performed with nothing to justify it, without a possible rationale. We find in Kuhn a combination of these two senses of faith.

Clearly Kuhn is after something important, quite likely true, in these allusions to faith, something positivists and mechanistically bent thinkers have simply ignored or actually denied. This is that people, including scientists, need to face some problems all alone, with nothing outside of themselves to fall back upon. Kuhn as well as Toulmin (1953) point to something like this in the lives of professional scientists. I shall say a good deal more about this in my positive remarks.

But as with many existentialists, Kuhn also omits a thorough examination of those types of circumstances. When we identify the nature of this 'act or leap of faith,' it could turn out that some very definite traits are required to produce satisfactory results (Koestler, 1964). It need not turn out that such an act necessarily excludes objectivity. But for now I am concerned with the conclusions Kuhn draws from his 'leap of faith' doctrine for purposes of understanding the concept of scientific truth, including whether we might not, after all, be able to speak of a correct or even true paradigm. These conclusions may now be turned upon Kuhn's own ideas.

I noted above that Kuhn uses the concept 'right' in connection with his own claims (1970: ix). In that context Kuhn is offering an objective conditional statement, one he believes could be true: if his theory is sound, then his statement is true.[12] So in this case, at least, Kuhn believes one *can* be right. And since he also says he is working here as a historian and sociologist of science (1970: 81), he allows objective knowledge to enter the realm of science in at least one instance.

Yet by its own meaning that conditional proposition denies to him the prospect of such objective truth, that is, knowledge of history. Consider what he says about what scientists can see and know:

> during revolutions scientists see new and different things when looking with familiar instruments in places they have looked before. It is rather as if the professional community had been suddenly transported to another planet ... (1970: 111)

Unless we confine the meaning of 'scientists see' to the strict sense of 'visually perceive' (which is not what Kuhn's usage indicates), Kuhn's own framework makes it impossible to speculate on his being *right* about anything in the above objective sense of the term.

Kuhn's problem, I submit, lies in the rejection of the possibility of objective truth in science – the idea, that is, that scientists can know reality. This rejection follows from an invalid conception of the nature of truth offered within the positivist/inductivist framework Kuhn is challenging. As that framework prescribes it, the idea is indefensible. But that does not prove the impossibility of scientific truth. The idea of what truth is must be supplemented by a clarification of the nature of meaning. J. L. Austin's advice must be heeded in this connection when he tells us that something is quite wrong if 'we suppose that language (or most language, language about real things) is "predictive" in such a way that the future can always prove it wrong. What the future can always do, is to make us revise our ideas…' (1961: 56–7).

The Contextual Nature of Truth

We can reconcile scientific truth with the fact that scientific theories undergo constant change in light of new circumstances if we adopt a contextual notion of truth. On such an account, one can say what is true, in or outside science, at any given stage of human knowledge. A true statement would be the best account of what is the case available at a given time.[13] Truth is a property of propositions, and the criterion of truth is the best and most complete correspondence with reality. But we can only determine whether propositions correspond with the reality we know, not with the reality we do not know.

Therefore, if truth is possible at all, it must be relative to the context of knowledge that exists at a given time. (And we need to remember that often what is taken to be true is just a very plausible idea, not one that has been as fully backed as possible under the circumstances.) New facts may require the revision of old truths. But as long as we are true as fully as possible to the facts that are available at a given time, we are entitled to claim that we have true propositions. If we were to regard all of the well-borne-out judgments of past sciences as false because in time they needed to be modified, upgraded, filled in, and so on, we would have to end up with the very odd conclusion that at no point do any of the sciences manage to produce any truth – after all, none of us is privileged to know that in the future these judgments will not require some modification, revision, and so on.[14]

Thus scientists can and often do know reality. It is philosophical ideas about truth that need to be called into question, not the capacities of scientists (Madden and Harre, 1973: 129). When one knows something, one must be prepared to show the following: (a) that what one claims to know is true, (b) that one can support one's claim in terms of the standards of truth appropriate to the subject matter. In the case of (a) we need to spell out what it means for a statement to be true in any particular context. Thus 'the surface of that wall is solid' is true if the wall has a solid surface *and* the statement is made in a context where it is clear, for example, that by 'surface' no one means 'the surface of a microscopic portion of the wall' or by 'solid' no one means 'capable of withstanding blows or other forms of penetration against all or any sized and shaped objects.' This is to make clear that statements are true contextually and that dropping the context, as Eddington did when he said that tables were not solid objects because there were spaces between subatomic particles, can change the truth value of the same statement (that is, uttered sentence). Also, we need to specify just how open-ended a concept can be, used in statements in different contexts. For example, to say 'the surface of that wall is solid' will be false if uttered by a house painter who is considering the wall for his purpose but true when said by someone who wishes to use the wall as a backboard for ball-bouncing.

These two points about truth need to be emphasized because often philosophers demand unreasonable standards of 'accuracy' or 'precision' in the use of concepts in statements which might be employed to say things they believe are false or meaningless. Any reductionist thesis, for instance, about what standards a statement must satisfy so that we may judge it true tends to move toward this kind of unreasonableness. Yet, without regard to the purpose and context of the discussion, it is entirely hopeless to attempt to specify standards. (In particular, empiricists try very hard to come up with some *ultimate datum* to which all true statements can be *reduced*; as it were, a true statement is thought to be capable of being true only about one kind of thing, for example impressions, sense data, and so on.)

In the case of (b) I want to indicate the difference between what putative knowledge claims and other assertions (for example hunches, opinions, and the like) require by way of justification or defense. If someone says, for example, 'I know Red China has ulterior motives for inviting the President of the United States,' we would require more evidence from him than if he just said he believed it.

What does all this imply about moral knowledge? First, that the standards for some statement to be true in a moral context could very well be different from others, just as the standards of the truth of a statement about physics are different from those made about mechanical engineering or poultry. And by 'standards' I mean 'specifications pertaining, for example, to the relationship(s) between *what is said*, the speaker, and what it is *about which something is said*, and so on.' Thus, for instance, '$E = mc^2$,' as an example of a statement in physics, must satisfy different standards of what will render it true from 'scales found on chicken feet indicate evolutionary relations to reptiles;' here what is said relates in different ways to what it is about which something is said: this is shown by *how* the investigators or scientists of the two fields *differ* in their method of checking out whether what is said is true.

If, therefore, there is moral knowledge, what we need to show is that there is an area or realm (of life or reality), just as with physics, chemistry, forestry, poultry, economics, and so on, to which moral discourse pertains, about which moral or ethical (and political) assertions are true, false, probably true, good hunches ('Maybe you ought to help him'), and so on, and relative to which one must go about justifying, supporting, assessing, and so on the things one says that one claims have to do with morality.

Our concepts of truth, knowledge, right, and the like have, for the most part, suffered utopian characterization by philosophers. Yet this does not mean that *people did and do not actually know reality,* that *true statements have not been uttered* by scientists in virtually any period of human history (even if within the corpus of a scientist's work there will most likely appear a mixture of true, partly true, and false claims). The idea that truth is impossible is absurd – it amounts to holding that all of science consists of claims that are neither true nor false.

But philosophers can make mistakes; they can have false ideas, including false ideas of the truth. Kuhn seems to be right enough in indicting the 'traditional' (positivist) philosophers of science for their characterization of truth. But instead of setting out a revised and better characterization, he denies the very possibility of objective truth.[15] (And by 'objective' I do not have in mind the narrow/impossible sense of that term used by absolute idealists, nor what is meant by positivists, behaviorists, and other empiricist-oriented thinkers, namely, roughly, 'capable of being supported by references to sensory experience.' I am using the idea to mean something closer to 'capable of being provided with a justification in terms of standards appropriate to the subject matter.') In his efforts to show the errors and misleading implications of positivism and empiricism, Kuhn goes on to deny something that is not at all strictly tied to the views he criticizes, namely the possibility of objectivity in any way, shape, or form.

The Rational and the Logical

Interestingly, however, Kuhn runs into difficulties beyond simply excluding his own position from the realm of objectively true statements and conclusions. He may try to extricate himself from the self-referential problem with the admission that 'right' (or 'true') means for him what *his* paradigm forces upon him, no more. And who are we to challenge his choice about how he *will* use familiar concepts? Everyone else is playing havoc with language, why can't Kuhn? A more severe problem arises for Kuhn when he adamantly distinguishes between being *unreasonable* and being *illogical* – in the case of Priestley, for instance (1970: 158).

According to Kuhn, Priestley was unreasonable but not illogical. This distinction is made even more explicit by Morton L. Schagrin when he tells us, in what amounts to a defense of at least some of Kuhn's main points, that 'Priestley was always rational … he was merely "unreasonable," not irrational' (1973: 9). The meaning of 'rational' and 'logical' in these passages is clearly 'loyal to the canons of sound reasoning, valid argumentation.' Kuhn, as well as Schagrin – who is actually closer in his views to Feyerabend than to Kuhn – would admit that these canons remain entirely indispensable throughout the history of *bona fide* science. Indeed, to deny this much to science would allow all manner of occultism to claim the status of scientific activity.[16]

Yet if we insist on logic, we cannot dispense with the idea of objective truth. The ideas of 'rational' and 'logical' in the above sense must be explicated in terms of the concept of validity. But validity, as I noted earlier, implies that if the premises of an argument are *true*, the conclusion must also be *true*. But this presupposes the meaningfulness of the concept 'truth.' The idea of *proving* some conclusion assumes that there exist premises that have truth value, that provide support for the conclusion. If it is impossible to obtain such premises, and if we cannot know whether such premises are true, then proof is meaningless. There is no point in insisting that scientists proceed *rationally* unless such a procedure *can* secure for them some measure of contact with reality.

Other things, such as experience, experiments, and instruments, may be required for scientific inquiries, but rationality as a minimum is indispensable. Kuhn seems to recognize this (Aristotelian) doctrine, as does Schagrin. But because of a commitment they have to a doctrine about the relationship between logic and reality – for example, a form of the analytic/synthetic dichotomy – they do not accept the presupposition of rationality, namely that human beings can know[17] that they have identified reality correctly.

I should make clear that the suggestion that false premises are sufficient for logic invites the reply that falsity is inseparable from truth. But the objection that to understand the meaning of the concept 'truth' does not require any knowledge whatever cannot be treated here. A full theory of concept formation and concept validation would be required (see Rand, 1966).

There are traces in Kuhn, and even more in Feyerabend (1963), of a full-blown conceptualism. The above points must suffice for now as crucial objections to this view, even if an adequate alternative cannot be offered here. If anything entitles someone to reject another's views, then the fact that those views deprive him of the possibility of even ascertaining the subject matter of his discourse (ideas, theories, discoveries, and so on) must suffice for that purpose.

In the end, despite many valuable and correct points, Kuhn's work leads to unacceptable conclusions. Kuhn disallows the possibility of knowing whether one's conclusions are true, whether one's beliefs are correct – in science or in any other field of concern to human beings. My critique of Kuhn concentrates on those features of his position with significant consequences for the theory of knowledge and the philosophy of science. Kuhn says that Descartes's philosophy is a paradigm (1970: 1, 12). Thus he must accept that his own position qualifies as an alternative paradigm. By showing that, when turned upon itself, his theory of paradigms leads to the impossibility of our ever learning if it is correct, that theory is seriously, if not fatally, undercut.

Notes

1 Although I take issue with many of the philosophical points in Kuhn's thinking, I should note that his historical reportage has been very helpful and mostly born out by the evidence. It is his interpretation of the history he reports so well that is my concern. The philosophical, especially epistemological, perspective he deploys is what I am focusing on here.

2 By 'meaningful' I do not mean only that we could *imagine* what it would be for those things to have transpired; I have in mind here rendering it possible to *conceive* of the actual events as having transpired as he says they have, that is, actually learning of what occurred. In history, also, we are not after imaginable episodes but after *knowledge* of what happened. Yet paradigm-induced images of the past cannot provide that, as Kuhn must admit. (Observe how he speaks of what 'Priestley was, to the end of his long life, unable to see…' (1970: 56). This is not a statement of historical fact but a Kuhnean paradigm-colored rendition of history by Kuhn's own doctrine. Are we justified in using it as evidence, then, in support of Kuhn's theory of paradigm determination? Not without obtaining a circular argument. Consider the evidence of misconduct among scientists presented in John Waller, *Einstein's Luck*, New York: Oxford University Press, 2003, wherein scientists are viewed more as moral agents than as mere captives of various perspectives that forced them to either see or be unable to see various things.)

3 I am aware of the controversy over the validity of circular arguments and I also appreciate that Kuhn cannot be asked to settle that controversy before proceeding with his case. But again I want to point up how dangerous it is to provide impossibility proofs when the bases are not sufficiently covered.

4 This follows from Kuhn's view that (a) choice of paradigm rests, in part, on an 'element of arbitrariness' and (b) philosophical positions, too, are paradigms or paradigm governed (1970: 4–5, 121). It should be noted, however, that to make this case one would have to argue first that all (including metaphysical and epistemological) truths or principles *must* be arbitrary. Yet the concept 'arbitrary' cannot be rendered intelligible if all of our knowledge rests on arbitrary standards since that concept is meaningful only in contrast to the possibility (that is, actual prospect) of some knowledge that is demonstrably or axiomatically (that is, undeniably) established.

5 I am thinking here of the rationale for the courtroom procedure involving the standards for cross-examination that only points raised in direct testimony can be brought into the latter.

6 Whether scholarship bears out the charge that Plato himself advanced this model, in view of his admiration for mathematical knowledge, is not important here. The point is that that model has been thought to guide efforts to provide an account of human knowledge. Rationalists as well as empiricists, up until recently, have found it necessary to characterize all knowledge in such a way that it would have to satisfy the requirements of that model. Only recently has it been emphasized that extrapolation from mathematical knowledge (or, alternatively, from the immediacy of sensory feelings) might be an unjustified restriction on what could count as knowledge of other aspects of reality.

7 If we allow that Wittgenstein was focusing here in part on his own earlier commitment to logical atomism, with its affinities with empiricism, the emphasis on perception can be appreciated. But this should also warn against extrapolating the results of this emphasis to other forms of awareness of reality.

8 For example, if we revise the notion of 'common features' to allow for other than observable or sensible aspects of what we want to classify, and if we allow for contextually determined modifications of the applicability of our concepts.

9 That is, with its emphasis on the impossibility of finding ultimate truths – something pragmatism shares now with several other philosophical schools, for example existentialism.

10 That is, in the possibility of *philosophical* knowledge!

11 In religion, for instance, one has faith not because nothing else is available to account for some features of reality and human life – that, as some theologians claim, would be too easy. Only when faith takes the effort to overcome even contrary evidence can it have its proper significance. This is not, however, typical in theism.

12 A note of caution here: the concept 'true' is used here (as it is in most discussions) to characterize – or to assign a certain property or attribute to – judgments, beliefs, statements and such. Those that are – that is, have this property or attribute of being – true correctly identify facts. But it isn't facts that are true, facts simply are, they are what is the case, what exists, what reality consists of. A discussion of truth is not the same as a discussion of facts. Whether to be true it is enough for a statement, judgment, proposition, belief to identify a fact depends on how rich the concept of 'identification' is in the analysis. Does identification involve being correct or justified identification? (That is, is 'identification' a success term, such as 'knowledge'?) These are issues I do not take up here.

13 Kuhn does not seem to think that an approach such as the one I am sketching can work (1970: 101–9, 136–43, *passim*). I suggest that Kuhn is emphasizing anomalies altogether too much and does not consider that they might or could have been avoided in at least some instances while those that remain would not, then, justify drawing Kuhn's conclusions about the *nature* of science. It could turn out that we just *do not* know enough to account for those instances.

14 The best prominent account of this conception of knowledge and truth may be found in J. L. Austin, 'Other Minds,' *Philosophical Papers* (Oxford: Clarendon Press, 1961).

15 The expression 'objective truth' is used to mean the theoretically distinctive characterization of what 'truth' must mean in terms of the objectivist approach. Outside a philosophical context 'truth' would mean just what such an approach argues it must mean, assuming the argument is itself sound. Saying truth is objective is something that has a point when discussing various ways truth has been characterized, accounted for. Otherwise it has no point – no other truth but objective truth is truth.

16 Kuhn and those sympathetic to his ideas would still hold that pervasive conceptual confusion within a field indicates the absence of at least the necessary conditions for something being a science. One can ask, however, whether Kuhn could justify even that much – why, that is, should we insist on logic? Some commentators on an earlier version of this discussion posed just that issue: why is the requirement of logic unquestioned herein? Certainly there have been those who have called this, however paradoxically, a 'Western superstition.' Indeed, I would argue that such a challenge is incoherent: there simply is no possible way to identify its meaning. Without logic it could mean anything whatever.

17 I suspect that I am offering an approach that owes more to features of Aristotelianism, with contemporary renditions and modifications, than to anything properly construed to be modern or progressive. Not being a historicist, I can rest at ease with the possibility that temporal development, even obvious accumulation of some kinds of knowledge, does not guarantee advance in all fields. I can imagine that even modern science owes more to earlier philosophy than to the philosophical views of its own times, at least when we consider the subtle fashion in which philosophical ideas have an impact on people's lives and ideas.

Chapter 4

Kuhn and the Moral Dimension of Objectivity

Introduction

In *The Structure of Scientific Revolutions*, Thomas Kuhn does not merely undermine the epistemological foundations of scientific objectivity, he undermines its moral foundations as well. Kuhn claims that the choices scientists make from among different, incommensurable but in some sense equally plausible paradigms are ultimately subjective, determined by such factors as 'personal and inarticulate aesthetic considerations.' Since we have little control over such personal factors, Kuhn implies that scientists do not act freely, but are moved by factors outside their own control (1970: 56, 75, *passim*) – at least in the crucial stages of their professional lives. (How they might still act freely at other stages is not clear from Kuhn.) If scientists are not in control of their minds, then the ideas they choose to adopt will correspond with reality – will be true – only by accident. Furthermore, if we understand objectivity as a conscious policy of pursuing objective truth, of basing one's beliefs on objective evidence rather than on subjective factors, then Kuhn implies that objectivity is impossible as well.

In this chapter, I wish to counter Kuhn's arguments and sketch an alternative approach to understanding the history of science, an approach that affirms the freedom and the moral responsibility of the scientist to pursue objective truth in a self-conscious, methodical manner.

Scientific versus Philosophical Perspectives on Man

To understand what acting as a scientist requires, we must first understand acting simply as a human being (Louch, 1966; Chein, 1972). Kuhn relies on sociology to tell him what science is. But this is problematic, because sociology is a science itself, and as such it is governed by standards laid down outside of the sciences. This is natural, since scientists are more likely to know about the objects of their studies than about the necessary conditions for studying anything at all. This is a question for epistemology, which is not one of the sciences.

Epistemology, however, is not the first or most basic branch of philosophy. Epistemology studies knowledge. But before there is knowledge, there must be something that is known. If nothing is, then there is no knowledge either, and epistemologists might as well quit while they're ahead. The philosophical study of what is, is metaphysics.[1]

Metaphysics, however, has not fared well since the time of Descartes, who oriented philosophy toward epistemological issues. Sociologists, guided by epistemologists, had plenty of guidance about what knowledge might be, but hardly any about what reality is. Method took precedence over content. The latter had to be constructed out of the requirements and limitations of the former. Kuhn is to be credited with illustrating just how elusive the world seems under the influence of modern philosophy.

Positivistic, behavioristic sociology, hoping to imitate the successes of the natural sciences, adopted their methods (Louch, 1966: 7). This resulted in an insistence upon extrospective, empirical observations, the covert adoption of a materialistic ontology, and the pursuit of causal laws governing the relations of events. Even exceptions, like Marx, built on this idea instead of challenging it at the roots. (Marx paid homage to empiricism but gave it a supposedly empirically derivable dialectical twist.) The consequence of this approach for the history, sociology, and philosophy of science is to treat scientists as if they were basically passive participants in the unfolding history of the social institution known as science. For exceptions, see Louch (1966), Polanyi (1969), and Agassi (1963).

But it could be ontologically correct to suppose that different causal laws might govern the behavior of entities with physical properties, chemical attributes, biological features, *and* human characteristics. This pluralistic ontology makes possible a non-reductionist conception of scientific method. It makes it impossible to dismiss freedom *a priori*. Human beings could be causes of some events.

I am not denying the possibility of a science of human affairs – or many sciences of human affairs. Economics, political science, psychology, sociology, and so on could all be scientific in the sense of seeking objective knowledge of different aspects of human existence by employing appropriate methods of investigation. It would, however, be dogmatic, not scientific, to *presuppose* that mankind were ontologically indistinct from the subject matter of other sciences, forcing the sciences of man into a natural-scientific framework instead of developing an appropriately humanistic approach.

When sociology, for example, imposes a materialist perspective on human affairs, then the sociology of science will tend to do the same, and room for genuine choice – a creative mental act – will be missing from the start. Kuhn the antipositivist falls prey to this, but for his slight variation, similar to Marx's revision of mechanistic materialism (Marx and Engels, 1968).

Freedom and Moral Responsibility in Science

Thus far I have attempted to clear the field for some constructive suggestions.

What I want to argue is that once the room for choice, that is, self-determined, self-initiated action, is cleared, we can ask whether human beings can choose at least some of what they do. I will not enter here the dispute about freedom versus determinism because I could not do it justice. It does seem to me absurd to deny that people can choose (however much this is circumscribed by nature), particularly with regard to their thoughts (Boyle et al., 1972; Jordan, 1969; Machan, 1974a; Machan, 2000).

Without freedom, much of what we do would be meaningless. Knowledge, understanding, valuation, judgment, inquiry, research, criticism, improvement, modification, revision, and so on, would lose their meaning. This is because all these presuppose that human beings can choose to address problems carefully or not, that they are able to be independent rather than tied to various prejudices.[2]

Be that as it may, let me for the moment simply explore what follows from the acknowledgment of human freedom of choice.

Let's assume that people can initiate some of their actions, including scientific activities like research, reasoning, the selection of explanatory schemas, and the like. If so, then one way to approach the history of science is to ask what scientists chose to do and why, as well as whether their choices were justified, reliable, objective, biased, irrational, hasty, stubborn, and so on. Kuhn might consider this approach to be importing a *preconceived view* of human nature into the study of history. Nonetheless, Kuhn should be the first to accept such a move, as he himself imposes his own paradigm on the science of the past. Thus we are left to determine which is the *better* framework to impose.

I am *not* suggesting that we judge what scientists did on the basis of what Agassi calls the most up-to-date textbook version of some field of study (1963). All I am suggesting is that we must have a clear conception of what human beings are, what they can and cannot do, as well as what they ought to do in general, before we can hope to provide a correct rendition of any part of human history of science. (Again, by 'correct' I do not mean 'final' but what is the only sensible rendition of that idea, namely 'rationally justified.')

If people can choose and are responsible for thinking and judging objectively, then knowing people's characters and the circumstances they face can throw light on what they did, how well or badly they acted, and what results their actions had (Louch, 1966).

While such suggestions may appear rather mundane, they help us to understand aspects of human history that are invisible to Kuhn. I am particularly interested in the possible force of a certain kind of *ad hominem* argument that might be invoked in the explanation of scientific conduct (Gerber, 1974). We often do not have sufficient knowledge of the details of someone's thinking, the evidence he had available, the precise context of knowledge enjoyed within his field, and so on. If we had to rely on such information alone we would be at a loss to explain his actions. (It seems that Kuhn chose to invoke 'elements of arbitrariness' when he came to coping with such cases.)

I am suggesting that instead of settling just for 'social pressures' or supposedly subjective aesthetic preferences to make sense of conduct in such cases, we ought to consider people's characters. In other branches of history, particularly biography, the concept of character is indispensable. And, since scientific activities are part of the biographies of scientists, the concept of character should be relevant here too. Scientific conduct is not immune from moral considerations – nor is science itself uninfluenced by human virtue.

Of course the resistance to such an approach is overwhelming. Several factors account for this reluctance. One is that science is supposed to deal with facts, whereas most commentators and serious scholars consign matters of moral character to opinion or superstition. But there is a great deal wrong with this dichotomy, and Kuhn is among those who have helped us realize that. Another source of resistance is the belief that explanations *must* be subsumed under general laws. Here again recent discussions in the philosophy of science point to this as either too simple or actually quite compatible with the idea that people are free and responsible (Louch, 1966; Scheffler, 1967; Madden, 1969; Machan, 1974a). Yet another source of resistance is the enormous difficulty of agreeing on what moral standards to apply to people in general, as well as to scientists as such. While Scheffler and some others have suggested an answer to what is morally central to being a working scientist (Scheffler, 1967; Polanyi, 1969; Skinner, 1976) the issue of what is central to morality as such is not easily approached – not to mention the fact that few agree that anything *could* be central in the first place (Winch, 1967; Beardsmore, 1969; Wheatley, 1969). Still, the fact that there is resistance to an idea does not constitute an argument against it.

Having covered so much ground so hastily, introducing additional tangents and elaborations would invite criticism on just that count alone. I will instead focus the rest of my comments and suggestions on the issue of the place of moral considerations in the history and sociology of science. These remarks assume the possibility of identifying a basic set of objective moral values which do not change from time to time and place to place.[3] The very possibility of a sociology of science requires an understanding of at least some of these values. Furthermore, there should be no *a priori* denial of free will and moral responsibility within science. If my suggestions are fruitful, these possibilities may turn out to be important realities within the history of science, and their acknowledgment could be necessary for understanding and explaining human behavior on all fronts.

Applying a Theory of Scientific Conduct

In this section I hope to provide initial grounds for employing a more individualist, albeit not atomistic or Hobbesean, approach to the history of science. One consequence of this approach is that periods of crisis in science can be seen as moral challenges to those who face them. We need to consider scientific events from the point of view of the participants. To do this, we have to reconstruct the situation in considerable detail. We then have to employ standards of conduct by which to evaluate the choices of the actors. Of greatest importance is the realization that those actors might have done otherwise.

Priestley, for example, despite his protestations, might have approached his problems with greater receptivity, perhaps by going beyond the issues that lay at the surface of the dispute with Lavoisier, beyond the actual scientific problems all the way to a consideration of the metaphysical underpinnings that his approach took for granted. For this purpose it would not be enough to examine Priestley's own letters and papers. The 'raw data' of history would not be sufficient to yield an understanding and explanation of what happened.

When Kuhn thinks that a 'different concept of science … can emerge from the historical record of the research activity' (1970: 1), he indicates his opposition to the approach I am suggesting. Neither for the general understanding of science nor for the more specific understanding of scientific episodes is it sufficient to expect answers to simply 'emerge from the historical record.'

A general skeptical objection to these suggestions is worth considering,[4] namely that we cannot really expect to determine the value of scientific conduct, especially as regards its fruitfulness in discoveries and explanations, by paying attention to the character of the scientists' behavior. This is because we can find people who make what turn out to be very good decisions while conducting themselves quite irresponsibly in the process of arriving at them. Such individuals have been known to exist in the history of science. The conclusion emerges that unjustified, even irrational, behavior could produce valuable, one might say entirely correct, results. Conversely, other entirely conscientious individuals and groups can be seen to have made only minimal progress in their endeavors. Surely in the light of such cases it is fruitless to invoke such putative determinants as the scientist's moral character.

Clearly, the *method* I am suggesting simply cannot *guarantee* that one who always lives up to standards of good personal scientific conduct *will* produce great scientific results. In all areas of human life some of the worst people succeed. Those who abuse their health may live a very long life; those who drive carelessly may make it to their destination; those who fall from second-story windows may survive the fall and even write a best-selling book about the lucky event. Luck is clearly a factor in many a human endeavor, science not excluded. In addition, it is not possible to know always just what steps *must* lead to a discovery. There are surprises, even anomalies, in most areas of human conduct. The most vicious killer can carry out his vice with commendable results – as when he accidentally guns down an even more vicious felon.

But to construe this as a failing of an explanatory scheme or approach requires that we entertain an impossible model of explanation. Kuhn is right that 'methodological directives, by themselves … [are insufficient] … to dictate a unique substantive conclusion to many sorts of scientific questions' (1970: 3), including to those raised in the history of science.[5] Only to those who choose to entertain the false alternative of ideal certainty or skepticism will this be a failing on the part of a theory.

Having taken note of what the suggested approach cannot and does not promise, let me however indicate why the objection is not quite as forceful as even I have appeared to allow. A careful investigation of the Priestley/ Lavoisier affair along lines suggested here will not rest with the observation that Priestley's obstinacy managed, after all, to produce a great deal of useful research, some of which led to important findings. In such cases historians tend to focus on the beneficial consequences of someone's admittedly 'unreasonable' conduct (Schagrin, 1973).

We could, however, approach the matter differently. What of the effort that *could have* been spared? If we admit that, despite protests, the stubborn men of science could have yielded what amounted to sound arguments in their context, that they could have gone a bit further in their inquiries to check their and the tradition's premises beyond what was conventionally appropriate, and so forth, then this line of inquiry will no longer appear hopeless. Notwithstanding Kuhn's protests over claims about 'what might have happened' (1970: 127), it is appropriate in our effort to make sense of human history to consider alternatives to the actual development of events.

What this implies is that even if we admit that reckless and improper behavior can produce valuable results, we can still ask whether it might not have been better to abide by the proper canons. By insisting that people could have done otherwise, had they chosen to abide by these standards, we undermine the force of the objection involving anomalies.

Let me now return to some of the possible benefits of the approach I have been proposing. One area in which we may find it helpful is in considering biographical data. The issues dealt with here can have a crucial bearing on what we will admit as determinants of scientific change. A historian who takes the appropriate moral point of view will, first of all, be able to draw on such data to develop a general idea of the moral aspects of the scientist's life. I do not mean his sexual habits or how often he wrote to his parents, of course. The sense of 'moral' I am using here goes to the more basic aspects of a person's way of living his life *qua* person – whether he has integrity, whether he has cultivated an honest approach to his own professional competence, how he treats the material upon which he speculates, how he compares his work to that of others who may be working in opposite directions, and considerations such as these.

Of course to gain this kind of understanding of his subject matter, a historian must at the same time have a clear grasp of what would constitute the scientist's actual professional universe. Just having a moral framework isn't enough to understand the actions of a scientist and explain the consequences of those actions. But the moral point of view is crucial for understanding the choices of scientists in certain circumstances.[6] Given that there are better and worse things that people can do with their lives, and that some people devote a good deal of their lifetime to science, I take it as given that there are better and worse things that people can do with their time in their profession also. The purposeless pursuit of bits and pieces of truth does not seem to me to be so commendable as it is for admirers of so-called 'pure' science.

These points simply give a clue to what I would do in a broad examination of the history of science. But they assume a certain moral perspective already, one for which I have offered no justification. The present thesis does not defend a particular moral code to be cultivated in science, only that we consider the issue seriously. That is because general philosophical considerations touching on the issue of human nature and conduct indicate that people can be free to choose some of what they do, and that some of what they do is better than alternatives. And this would seem to be true of all areas of conduct, science not excluded. The historian of science would then be well advised to approach his studies in the light in which human conduct in general needs to be considered. While ordinarily relying on moral considerations for understanding the history of science may not be necessary, understanding and explaining the choices and actions of scientists in times of crisis may require them.

Let me now turn to such 'revolutionary' scientific changes, Kuhn's central focus of interest. First we will have to consider just what Kuhn takes these eras to consist of and what makes them revolutionary. What exactly undergoes these revolutions? Kuhn tells us that 'during revolutions scientists see new and different things when looking with familiar instruments in places they have looked before' (1970: 111). For example, Kuhn argues that, 'Pendulums [not the idea of 'Pendulum'] were brought into existence by something very like a paradigm-induced gestalt switch' (1970: 120). Kuhn might well mean this literally, given his position and its implications.

The alternative I want to suggest will help us in understanding whether Kuhn's points could be advanced in a way that does not result in insurmountable problems of subjectivism. No doubt the *concept* of 'pendulum' may have been introduced to identify a phenomenon that had thus been rendered more sensible by Oresme and, later, by Galileo. But is it correct to talk of the 'invention' of a paradigm? And thus of the invention of the pendulum? It is philosophically more accurate to speak of the discovery or identification or acknowledgment of a more meaningful, more sensible, more coherent, more consistent, more parsimonious conception of reality by the formulation of a concept amply supported by both rational and empirical considerations. Even if those who introduced it had only a 'hunch' about the need for this new concept, it makes a shambles out of science to talk of inventions, to extrapolate from the language of technology, of design, of engineering.

Not that Kuhn has no insight here. But akin to certain philosophically eager scientists (for example, B. F. Skinner, W. Heisenberg, A. Eddington) Kuhn extrapolates from an area where the concepts of creation and invention apply, for example in art and technology, into one where they have only questionable functions (that is, science). Kuhn admits that he 'cannot yet specify in any detail the consequences of [his] alternative view of scientific advance' (1970: 171). Despite this and other cautionary remarks, however, Kuhn's general proposals are far too firm, considering the shaky foundations of his approach.

Among many valuable features of Kuhn's book, he points to an important feature of concept formation in the development of human knowledge, and he is making an important departure from the dominant view of science when he proposes that, 'If we can learn to substitute evolution from-what-we-do-know for evolution toward-what-we-wish-to-know, a number of vexing problems may vanish in the process' (1970: 171). Yet he undercuts this suggestion with his relativist metaphysics in which the very possibility of scientific knowledge has no place, and in terms of which not just knowledge but the facts, entities, relations – that is, existence itself – change during revolutions. We have seen Kuhn defend this metaphysics by way of some extrapolations from psychology and the history of science (as he interprets it). Another aid Kuhn invokes is the idea of political revolution (1970: 92, 160).

Concerning this last piece of ammunition Kuhn fails to consider a crucial point about political knowledge. Such knowledge purports to be about what people *ought to* do for purposes of organizing a good and viable human community. One need not embrace the is–ought dichotomy to acknowledge that the 'is' of political theories does not function in saying what people actually do in relation to politics; the 'is' in such areas functions to point out what is right and good.

Thus the reference to political revolutions can be very misleading. The subject matter of political theories purports to be the political principles people *should* accept and practice. Only revolutions in political *science* could serve to satisfy Kuhn's illustrative purposes. Do political revolutions actually change *what should* be pursued within the context of organized human communities – that is, *is* constitutional democracy the best, most sensible account of how a human community should be organized *only after the revolution* whose victors advocate it? Is socialism how human communities ought to be organized only after the Bolshevik revolution? Only if that is the case will the example of political revolutionary change give support to Kuhn's thesis. But it is far from well established that such a Nietzschean approach to values is valid. Kuhn's analogy begs the question of his own ideas.

And this invites possible alternatives. First of all, Kuhn's schema fails to account for the possibility that we might be able to identify alternative choices scientists could have made – what they might have done, what might have developed, what advances might have been made, were it not for their actual choices. Kuhn uses the language of hard determinism throughout his work; for him scientists *had to* advance their findings as they did, they could not see what they did not see, they had no other alternative than to do what they in fact did (1970: 56, *passim*).

But there is reason to believe, based on a general philosophical theory of human action (Louch, 1966; Chein, 1972; Yolton, 1973), that scientists could have done otherwise than they did. And this suggests that some of them probably permitted themselves to advance theories they had not fully checked (by reference to a broader framework of understanding reality); some might have been stubborn, harbored irrational fears of controversy, lacked the courage to speak up when they should have, yielded to fads, failed to check on reports given to them by eager researchers, concocted research projects not warranted by what they knew, acquiesced (avoidably) to those not tolerant of new developments in their field, and so on. All these and other possible issues can arise, despite Kuhn's protestation about the difficulties with 'what might have happened' statements (1970: 127).

Second, Kuhn's account allows that in science those in the ascendant call themselves normal, established, respected, and trustworthy scientists, while those at the outskirts – even outright irrationalists – characterize themselves as the wave of the righteous future. Only social pressures, which can clearly be wrongheaded, even irrational, operate to maintain the usefulness of critical concepts, of a semblance of orderliness. By adopting the more cumbersome, difficult, not at present very promising road of focusing on what *might* have been done by scientists, these paradoxical consequences could be avoided – without falling prey to the pitfalls of a positivistic analysis. Kuhn's historicist (conventionalist, dialectical) account isn't the only alternative.

Third, in human affairs individuals must at times be trusted or condemned not on grounds of what they have done, but on grounds of what kind of human beings they are. *Ad hominem* arguments are attacked in logic textbooks, yet character references are used in courts of law, where drastic measures are taken with their aid. Gerber puts the point well:

> By using some abusive, neutral, or laudative terminology one can say something which implies that a person fails to tell the truth. One can even say something which implies of a man the chronic failure to tell the truth. One can call someone a liar, a cheat, a fraud, a swindler, a humbug, a quack, a charlatan, and a faker, using strictly abusive language, as well as a storyteller, a perjurer, and a pretender, using language which may on particular occasions be abusive, and which on others may be neutral or even laudative. Each term on either list in its own distinctive way sometimes can be used to indicate that the person spoken of does not tell the truth at all, or does not tell the truth with regard to some special matter. In calling Jones a faker (a man who always fakes) it is implied that when Jones complains of a throbbing migraine he is lying, from which it follows that what Jones says in this case is false. (Gerber, 1974: 27)

Similar points can be made about arguments from authority, a form of argument clearly useful and capable of yielding not only valid but sound conclusions in our dealings with experts in many fields (for example, medicine, education, engineering, auto-mechanics, and science). I am not recommending, of course, that we invoke such arguments indiscriminately. But trusting others' judgments in cases where it is warranted can be a means to a rational account of events, *even* in science. In view of the enormous difficulty of gathering sufficient data in the history of science, as in trials concerned with establishing what must or might have happened, these sorts of arguments can be useful for purposes of yielding explanations of scientific conduct, even progress or regress.

For example, historians may find that they have insufficient knowledge of a period of scientific change to explain the choices made by scientists; or they may find that even detailed autobiographical reports indicate confusion, lack of sufficient support for one or another alternative approach to the handling of some problem. In short, the historian is at times unable to unearth the (possible) reasons that have led to *success*, that is, to the eventual rational justification of the alternative chosen. (Consensus is *not* the valid standard for evaluating success in fields where *knowledge* is the goal.[7]) Then and only then can such arguments as I have mentioned above serve the goal of explaining human action.

It is at these points we can invoke Louch's idea that, 'In appealing to reasons for acting, motives, purposes, intentions, desires and their cognates, which occur in both ordinary and technical discussions of human doings, we exhibit an action in the light of circumstances that are taken to entitle or warrant a person to act as he does' (1966: 4). In the less than technical but not quite ordinary area of detective work, a hunch, for instance, may be a good reason for someone's action, just in case we know enough of the person to testify to his 'feeling' for the relevant issues, even when he himself either will not or cannot give adequate support for that hunch.

Kuhn might find these sorts of considerations inadequate for explanation because, as we have already noted, he complains of the 'insufficiency of methodological directives, by themselves, to dictate a unique substantive conclusion to many sorts of scientific questions' (1962: 3). This is to set an unrealizable, thus incoherent, standard for what is to count as rational and reasoned scientific conduct and judgment. The 'God's eye view' of reason, rationality, and objectivity needs to be abandoned, not the view that reason should guide judgment and conduct.

In this connection consider Toulmin's point about Einstein's idea of 'physical theories as "free products" of the human imagination.' Toulmin admits that scientific discoveries are not inductive inferences, nor conclusions based on strict deduction; but he says that 'we must not be tempted to go too far. This is not the work for untutored imagination. It may be an art, but it is one whose exercise requires a stiff training' (1953: 29). Yet in studying the methods of the history of science, we are entitled not to focus on such cases. Kuhn fits the case well captured by Louch: 'The extreme case, employed to make the contrasting point, is taken as the standard or universal case, and thus the condition for any inquiry whatever' (1966: 7).

I cannot claim that Toulmin's reference to 'stiff training' and 'untutored imagination' mean for him morally significant aspects of a person. I do want to urge that we consider such references in the light of man's moral nature (the exceptions being actions under coercion or physical limitations). Is it not appropriate to require that someone in a field of science should undertake such training? (Quite wrongly, many think that morality pertains only to cases of human *interaction*, that it rests primarily on *duties to others*, which must subsume other moral notions such as responsibility, integrity, honor, and similar *virtues*. We are in an era where philosophers tend to opt for *rule* moralities.)

An understanding of the development of science can utilize the knowledge we have of a scientist's moral character. Such knowledge can aid us in our efforts to make sense of what someone (and some group) did in some particular circumstance. It may help one to understand also why others interacted with him as they did when faced with his judgment and conduct. It is unnecessary to conceive of this approach as imposing on scientists of the past virtues that could only be developed and practiced in eras other than theirs, for example those of the historian. Rational reconstruction must take heed of moral context!

I agree with Joseph Agassi about the advisability of a contextual approach to understanding the history of science, in preference to both the inductivist and the conventionalist alternatives. Kuhn's position is a variety of the latter. Yet even Agassi says something in this connection that lends support to conventionalists. He tells us that 'The rise of an idea is the outcome of a work of genius, an unaccountable development' (Agassi, 1963: 55). Joining Koestler (1964), Agassi (1963: 26–7) is invoking a notion of being 'accountable' that he has found wanting in connection with the Marxist view of the generation of scientific interests. 'Accountable' need not mean 'caused by factors.' Not, at least, when the factors *must* be external. We can account for the rise of an idea by reference to the moral character of an individual, by his lifelong devotion to his task, by his commitment to solving problems, seeking the truth. This is surely one way to talk about Socrates. Why is it barred out of court when talking about Galileo or Einstein?

Nor need the emergence of an idea be unaccountable in the sense that precludes the prospect of understanding the subtle rationale for it of which the scientist himself may not be aware. Why should a scientist be worried about how the idea emerged? He is concerned with the problem it may help him to solve.

Conclusion

It has been my aim here to criticize and offer an alternative to Kuhn's approach to the understanding of the history of science. I have suggested that a clear understanding of ethics would aid us in an attempt to understand the dynamics of scientific development in both the individual scientist and the history of science. Scientists who are not concerned with being objective are negligent and irresponsible. Kuhn, however, bars us from objectivity in principle, not only because of his relativistic epistemology, but because of his implicit acceptance of a determinist, materialist ontology, which denies freedom and moral responsibility – including the freedom to think objectively and the responsibility to do so.

Notes

1 Kuhn acknowledges the influence and place of metaphysicians but he is not concerned with whether some positions in metaphysics, that is, whether some metaphysical claims are true or false. He is making an attempt here to take a neutral or 'God's eye' position, as if he could escape the requirement to deal with whether his own metaphysical presuppositions are correct. Yet in the sort of discussion Kuhn carries on that requirement is central (1970: 41).

2 I address this issue at length in Tibor R. Machan, *Initiative – Human Agency and Society* (Stanford, CA: Hoover Institution Press, 2000).

3 That is, some value(s) may be axiomatic, rationally undeniable, while others may be contextually dependent.

4 I have encountered this point in many circles, in many different types of discussions, including in connection with ethics, political philosophy, and, of course, epistemology. In reference to the present topic it was brought to my attention by Professor Morton L. Schagrin, whose knowledge of the history of science is far more extensive than mine. Yet the point is actually more of a philosophical than historical one, dealing as it does with the problem of how borderline cases, anomalies, problems, and the like, have to be faced by someone who offers an account of some aspect of reality. In ethics, for instance, must an ethical theory be capable of handling any and all so-called 'desert island' cases? For a more detailed treatment of this matter, see Machan (1974b).

5 For a full discussion of whether rationality requires that methodological directives produce 'unique substantive conclusions', see Wittgenstein (1969) and Stroud (1966). I should like to suggest that it is people who produce those conclusions, not directives. But that does not deprive us of objectivity; indeed, as I read Stroud, it renders it our responsibility to strive for it. See also Scheffler (1967).

6 Incidentally, these remarks should not be taken to endorse moral supervision by governments. As I see it, a decision having to do with science, research, funding, and the many related matters really should not be a political one in the first place. See T. R. Machan (ed.), *Liberty and Research Development* (Stanford, CA: Hoover Institution Press, 2002).

7 Kuhn seems to speak of consensus without explaining whether it is the result of scientists having a common purpose, more or less, or because they all want to agree with each other. Keith Lehrer has offered this distinction as a very important one in his 'Social Consensus and Rational Agnoilogy', *Synthese* (1975): 141–60. Here again it would appear that the considerations that enter pertain in large measure to a scientist's character, or morality.

PART III
Everyday Life

Chapter 5

The Nature of Objective Moral Claims*

Introduction

Could such claims as 'George ought to feed his child,' 'Tyrannies are evil,' and perhaps even 'Mozart's compositions are more beautiful than Schoenberg's' be objectively true? Or are they merely expressions of our subjective preferences?

Objective claims supposedly rest on what we become aware of 'out there.' Such claims supposedly mean a feature of reality and not what we think or feel or desire about it or what our minds supposedly construct, based on, say, elements of our culture, ethnicity, national origins, or the like.

Yet, 'objective' is not identical with 'intrinsic.' A value judgment can be objective if it identifies a real relationship between something that can be benefited and something that is of benefit to it, without it being the case that the latter has some property of being good in itself, of having value all on its own. This is the intrinsic view of values and is one of several ways in which objectivity might be secured for value judgments. For example, there are those who claim that nature or a sunset or precious gem has value in and of itself, independently of being of value to something or someone. If this were the case, the claim that these are valuable things would be objective.

All objective claims rest on facts of reality, but such facts could be entirely relational. Nutrients are of value to living things, which would be an objective truth, but they may have no value at all in themselves, so their value would not be intrinsic.[1]

With reference to ethics, in particular, by 'objective' we can have in mind the sort of principle of conduct that applies to an agent in virtue of that agent's natural goals or objectives – goals or objectives that are themselves demonstrably valuable to anything of its kind. This can best be appreciated from considering an objective claim in medicine, wherein everyone works for the value of health. Or in engineering, where they all aim at functional efficiency.

A subjective claim, in turn, is dependent on the subject who advances it; it is one arising from a person's unique point of view, state of mind, or set of personal (or cultural or ethnic) preferences. A subjective claim is held to be 'true' just in case the subject who advances it thinks or believes it to be so.[2] There is no requirement for the claim to be established independently of such unique tastes and preferences although there can be some complications about this when it comes to ethics and other normative concerns.

The objective justification of normative claims is a perennial philosophical problem.[3] The difficulty of answering this question has led some disappointed people to subjectivism. In this chapter I shall set aside questions concerning aesthetic norms and focus solely on moral ones, with the understanding that some of what I argue will also apply to political claims.[4]

In the popular culture, and even in some social sciences, especially economics, subjectivism is the accepted orthodoxy. Milton Friedman states, for example, that 'of course, "bad" and "good" people may be the same people, depending on who is judging them.'[5] Rorty tells us that we 'cannot say that democratic institutions reflect a moral reality and that tyrannical regimes do not reflect one, that tyrannies get something wrong that democratic societies get right.'[6] These and similar subjectivist views concerning what is good and bad, right and wrong, are prominent in many fields.

It is worth noting here, even if briefly, that meta-ethical subjectivism is often linked with classical liberal politics. The reasoning goes as follows: if no one can know that something is the right thing for one to do, no one has any justification for coercing another to do it – apart from the idiosyncratic 'reason' that one wants to do it. But this is not a very promising argument. As Richard Tuck notes,

> It is common nowadays for people to say that moral relativism should lead to a kind of liberal pluralism: that (say) the waning of religious dogmatism paved the way for modern religious toleration. But Hobbes's work illustrates that there is no reason why this should be so. Moral relativism, thought through properly, might lead instead to the Leviathan; and the Leviathan, while it will destroy older intolerances, may replace them by newer ones.[7]

If all that we have on which to rest our value judgments are our subjective desires, it is true enough that no one can claim to be right about them. Yet it is also true that no one can claim that the invasion of another person, based on one's desires, is wrong. So the classical liberal position cannot rest on such subjectivism.

In any case, most of us do make normative claims and show confidence in our claims.[8] We criticize conduct by politicians, parents, teachers, police officers, and other thinkers when they do not square with our value judgments. We support policies, institutions, organizations, ways of life, and so forth, believing these to be right. We act as if we think that our views are correct and defend them, as best as we can, when we are challenged.

For example, there is near universal agreement that parents ought to rear their children to ready them for adulthood. The same goes for the view that life-preserving actions are superior to life-destroying ones, at least in non-extraordinary circumstances. It is also generally thought that one ought to stay out of the way of angry beasts.[9] (I hedge a bit only to make room for cases where someone refuses to assent to the truth of such claims because he or she wishes to disguise a failing, air a radical possibility to the contrary, or is defending a contrarian philosophical position.[10]) In the main, however, such claims are treated as if they were true in most circumstances, and even the exceptions are subject to a great deal of agreement. This behavior indicates that we think our value judgments are objective, not subjective.

Objectivity and Agent Relativity

I submit that the subjectivity or cultural and similar relativity of ethical and political claims is apparent, not real. Ethical claims are often thought to be subjective because they pertain to how individual human beings ought to act and that, in turn, depends to a considerable extent on who these individuals are and their particular circumstances. We cannot always expect what is objective to also be universally applicable. This is analogous to the way medical claims may appear to be, but are not, subjective. Although particular diagnoses often apply to people in terms of their special or even unique situations, some basic claims concerning human health are universalizable.[11]

It is crucial to note that there need be nothing universal about a judgment that is objective, in the sense that the judgment must apply throughout the class of human beings. Such a claim can be objectively true and yet apply to just certain limited cases or even just one case. The claim that a particular hat fits Harry, for example, is true, objectively, without its being true that it fits everyone. That George ought to write to his mother on her birthday may be objectively true but it need not be true that everyone ought to do the same.

Of course, in the case of the hat that fits Harry but not others, the means by which we show this – for example, sizing Harry's head and then sizing the hat – would have to be generalizable, so that when we say that the other hat fits Joe, the means by which we show this to be true are none other than those we used to show the claim about Harry's hat.

Yet the same approach is not granted when it comes to making claims as to what Harry ought or ought not to do. With claims of that sort it is often suggested that they are subjective and 'true for' Harry but not 'true for' others, not only in the sense that the claims' truth pertains to what Harry ought to do but also in the sense that whether Harry ought to do such and such is something that is only 'true' if Harry takes it to be such. Such claims are thus said to be not objectively but only subjectively true.

This suggests that the objectivity available in other areas – for example, in science, engineering, medicine, and in ordinary factual concerns – is not available when it comes to ethical claims. Yet it is evident that such claims are in fact akin to the sort of claims about a hat fitting Harry. There is at least one respect in which they are, namely, regarding their agent- or wearer-relativity. The fittingness of the hat is related to the hat wearer's particular head size.

The hat always fits some potential hat wearer. And whether it does is either true or false. What is universal is only that if it is true, then anyone with normal faculties and access to the relevant information can know that it is true. This is sometimes obscured by saying 'That the hat fits is true only for Harry,' as if the truth of the claim rather than the fittingness of the hat were something particular to Harry. And the same happens with claims about what someone ought to do. It may be that only one person ought to do something, but the fact that he ought to do it is true for everyone.

Objectivity in this context resides in or is related to the means of demonstrating, proving, or grounding some claim, not in the range of its applicability. Via such a proof, if we can obtain it, we will establish the claims we make as ones that are about the world.

That is the point of stressing objectivity, to note the kind of connection that exists between what we think and what is the case. Indeed, we note this when we admonish scientists or jurors or judges at the Olympic Games to be objective – to stick to evidence and sound reasoning, to avoid letting their feelings or wishes influence what they think about matters of fact, to rely on their sound judgments. The issue of universality comes into play only in that the claim that is supposedly objective could be established to the satisfaction of anyone who can understand the standard.

As an example of a reason why moral claims could be both objective and seem subjective, note that ethical claims do not identify facts that exist entirely independent of persons. 'Harry ought to write to his mother' is pertinent when the individual who is Harry exists and could do just that, not at any other time. But to call this subjective implies that the situation can be entirely reduced to the contents of Harry's mind. But clearly this is not so. Instead, the situation involves Harry's relationship to objective facts: the existence of his mother, their temporal proximity to her birthday, the personal and social significance of her birthday, the goodness of his obligations to his mother, the existence of pens and paper and postal services, and so on. These are all objective facts, yet they pertain only to Harry. Hence it makes sense to describe Harry's moral situation as both objective and agent-relative. And it is agent-relative because something must be of value to somebody or something, akin to how a distance is always *from* or *to* something.

Some, like G. E. Moore, dispute this, arguing that things can be good or bad without being good for or bad for anything. For example, the existence of the earth may be simply good, not as it relates to the living things on it. Yet this is wrong. The earth is good because it supports life. Without that, whether there is or is no earth – just as whether there is or is no Mars – is of no consequence.[12]

Raimond Gaita makes the following observation pertinent to this discussion:

> If I am deliberating about which is the best route off the mountain and I fail to arrive at an answer, I can pass the problem over to my partner; it is only accidentally my problem. If I am deliberating about what, morally, to do, then I cannot pass my problem over to anyone else: it is non-accidentally and inescapably mine.[13]

Should I receive advice from a wise person, I still must be the one to decide whether to take it. So it will ultimately fall on my shoulders how I act, including that I took or rejected the wise person's advice.[14]

Yet does this disturb the objectivity of the issue at hand? If I concluded that I ought to abandon my friend and seek my own route when we are both lost on a treacherous mountain, is there no objective way for me to tell whether this is right, whether the claim that I should so proceed is true?

The objectivity is not disturbed in the slightest by the inescapable personal element of the judgment, namely, that what I ought to do is intimately linked to the *I*, to who and what I am, something that is ultimately irreducibly individual. Indeed, it is because I am in the position to make sound judgments that objectivity is possible.

Of course, it could turn out that I address the problem on a substantially subjective level, based on my unexamined fears, prejudices, and preferences. I could be concerned about my health but mainly because I am a hypochondriac, not because of how health relates to my flourishing as a human individual in this world. Or I could want to be educated not because of how that advances my life as the human individual I am but because of my feeling that education will make me impressive to people.

The substantially subjective approach contrasts with one based on what is actually important to me, using as criteria what kind of being I am and what is required for my flourishing. This is what would establish, objectively, what is right for me to do, how I ought to live my life. Subjectivity would enter from my disregard for or inattention to the important facts and from my letting less important matters guide my thinking. As in the case of a jury's refusal to pay attention to the evidence linking the defendant to a crime rather than to the defendant's looks, tastes in music, and preferences for certain sports, the decision to do something would be subjective if immaterial, irrelevant factors were to influence it.

The Space for Subjectivity

So what appears to be subjective could, upon closer inspection, turn out to be objective. Yet it needs to be noted that a subjective or purely optional element can enter basically objective moral decisions. We may have an admittedly fleeting, even trivial, taste for something. But if it is rational for us to indulge such tastes – since life is better with some pleasures – then such an element does not deprive the conclusion about what we ought to do of its objectivity. All it does is open up several options as to what, objectively, one ought to do.

Objectivity does not imply uniformity. It could be the case that one ought to choose any of a number of courses of conduct, with any one being objectively right, if they do not differ from one another so far as morality is concerned.[15] Thus it could be objectively true that 'I ought to perform either a Mozart, a Liszt, or a Brahms piece during my concert,' while it is entirely optional and, perhaps, in this limited respect, subjective, which of these I perform.

Some would characterize as 'subjective' those factors that determine which of these pieces I will perform. Yet to characterize what one ought basically to do as determined subjectively would be a mistake, for it could be true that one ought, objectively, to perform at least one of them. Or, to put it differently, one may objectively have the moral responsibility to honor a promise to give a piano recital but it will be a matter of one's subjective preference what music one will perform. So it would probably be best to say that although there are objective determinants of what we ought to do, there are many optional – subjective – matters that will serve to determine our actual morally justified conduct.

Objectivity in Ethics

But how to discover what one objectively ought to do? I want to develop a neo-Aristotelian approach here and defend it along common-sense rather than extremely technical lines.

When we judge the goodness of something, in the last analysis we come down to ascertaining how it accords with its nature, with what it must be to be the kind of thing it is. Sometimes this is very indirect, as when some outcome is being judged on the assumption that the intention to achieve it is itself good. For example, we take it that winning a game is good for our team; we are here silently and often preposterously assuming that our team deserves the win, or something akin to that.

More important here is to note that the nature of something need not be, as Aristotle thought, some being within a thing, something in addition to the being the nature of which is at issue, let alone what Plato seems to have held, some separate intellectual object. The nature of something is, rather, the most rational categorization of the being in question, in line with the properties our cognitive faculties have identified and organized in the world.

Accordingly, a good apple more fully or better realizes the nature of an apple than a bad apple. A good apple is more real of – or more fully – *an apple* than a bad apple. This isn't to say, however, that a bad apple isn't an apple, only it isn't to as full or complete an extent an apple as a good one.

So it is with a tennis game or philosophy paper or redwood tree or beaver, and that is just what we see from how sports commentators, professors of philosophy, botanists, or zoologist go about appraising instances of the kind of things they tend to appraise. A game played or paper done by a novice is a genuine article but not a good specimen.

In the same way, a good human being is a fully realized, fully actualized human being. A bad human being is an unactualized, unrealized, botched, or corrupted specimen of humanity. The actualization and flourishing of the body is health. Its opposite is sickness. The actualization and flourishing of the 'soul,' one's character, is moral virtue. Its opposite is moral vice. These are, no doubt, controversial points and, like so many others under discussion here, have been debated at great length in the discipline of moral philosophy. My point here is to indicate only that the present approach to ethics and other normative areas sees objectivity just as available in these more controversial arcas as it is taken to be in the more mundane ones of medicine, botany, zoology. No doubt the details are far more complicated in ethics than they are in the various life sciences. The proper caring for one's life as a human individual is, however, comparable, formally, to the caring for flowers, animals, and one's own health.

Now some argue that we are not, in fact, distinctively rational and that reason, *à la* Rousseau and some others, is actually a perversion of our nature. My point here isn't so much to debate what exactly is human nature, although I do not find these objections telling, but to explain that goodness relates to our nature. (Those like Rousseau illustrate this when they criticize elements of human life and society that aim to realize the rational aspect of our nature. This way they show that they, too, understand 'good' by reference to a conception of human nature.)

Assuming here without additional argument that we are the only rational beings known to us, it is objectively true that we ought to think well. This is not to say we all ought to be intellectuals. Nor does it mean that we cannot entertain ourselves with fantasies or purely physical activities or take a nap when our brains are tired. We cannot think all the time, so any obligation to do so violates the 'ought implies can' principle.

The definition of 'human being' as the 'rational animal' is the best, given that all of our distinctive activities spring from this rational faculty. In short, this definition has the greatest explanatory power. It even helps us understand what makes possible the debate about what human nature is. The good life for a human being, therefore, is a rational life, a life in which reason is applied to the problems of living.

Work is one of the most important ways in which we apply rationality to the problems of living, so let us examine the role of objectivity in the choice of a career in order to illustrate how reason can be applied to the problems of living in general. First, we have to recognize that each human being by nature has a distinct temperament and set of talents and inclinations. Second, we must recognize that if we choose a career that accords with our nature, we will take pleasure in it. If we choose a career that does not accord with our nature, then we will be unhappy. Third, we must examine ourselves and come to a correct understanding of our nature. Fourth, we must find the form of work that best accords with it. This often requires painful expenses or compromises if one's best choice is insufficiently remunerative or simply impossible in one's social setting. Fifth, we must take the necessary steps to pursue our chosen employment. This entire process consists of judgments as to what we ought to do. These judgments or claims are objectively supported by humanly, but not always *publicly*, accessible facts about ourselves.

What would one say to a person who chooses the career of a killer, justifying himself on the premise, 'I gotta be me'? Such a choice is foreclosed by using the criterion of one's humanity – the kind of 'I' in question – specifically the concept of natural human rights, which governs our actions in a social setting. Taking another's life is a violation of these rights. One's choices are also circumscribed by other virtues deriving from different aspects of one's humanity, such as one's role as a parent, friend, citizen, or professional in medicine, business, or law.[16]

The Nature of Moral Imperatives

On this account, moral imperatives are 'hypothetical' rather than, as Immanuel Kant and Kantians would argue, 'categorical.' They take the form, 'If one wants to do X, then one ought to do Y.' They do not take the form, 'One ought to do Y no matter what.' In ethics the point is that life will be lived best in this or that way. That is, if we adhere to certain guidelines, our lives will come off most successfully as human lives. These guidelines are constructed by generalizing from experience.[17]

Accordingly, if it is true, for example, that 'good' means 'life-enhancing for the agent, *qua* the nature of the agent,' so that, say, it is good for plants to gain sunlight because sunlight enhances plant life, and if 'ethically' means 'chosen by the agent because it is to one's good as the kind of being one is,' then it can be argued objectively (not necessarily deductively) that one ought to choose to do what enhances one's life as the kind of being one is, that is, given one's nature. Given human nature, then, one ought, first and foremost, choose to think and to act on the results. This will most likely produce a successful life.[18]

Is–Ought and the Naturalist Fallacy

Of course we now run up against an ancient obstacle. My thesis appears to conflict with the still widely embraced 'is–ought' gap doctrine, attribute to Hume.[19] It should be noted right away that David Hume argued that judgments of what ought to be done may not be deduced from judgments of what is the case. Most take him to have argued, however, that such judgments cannot be derived from ones concerning what is the case, yet Hume produced innumerable such judgments from his study of history, custom, tradition, and human psychology. It is very doubtful, then, that Hume was a radical non-cognitivist, along the lines of A. J. Ayer and other logical positivists, although that is how many see him.

Without entering the controversy in full, let me just note that the belief that one cannot *derive* true claims as to what one ought to do from true claims about what is rests on certain mistakes. It mistakenly assumes that all derivations must be deductive arguments at every step of the reasoning process. Indeed, many forget that Hume also rejected deductive reasoning in the sciences, for the same reason he adduced when he did so *vis-à-vis* moral arguments. But quite rightly very few philosophers assumed he meant by that that scientific reasoning is non-cognitive!

The skeptical implications of thinking that no 'ought' can be derived from 'is' are supplemented by G. E. Moore's naturalistic fallacy, namely, the view that a sound definition cannot be sensibly questioned. The Moorean approach has it that because when one defines 'good' as, say, 'the greatest happiness of the greatest number' or 'human flourishing in accordance with human nature,' one is always justified in asking 'But, is this really so?'. This shows that in fact no such definition could possibly be right. A definition is necessarily true – or at any rate, connotatively transparent – so it cannot reasonably be questioned because if one looks at it, it must strike one as true.[20]

This Moorean approach, however, presupposes correct definitions to be Platonic, fixed, and impossible to conceive to be false. But definitions aren't akin to tautologies or deductively true statements. They are more akin to sound theories which manage to handle a set of problems most parsimoniously, most effectively and usefully. (Here we see another element of the misguided way knowledge has been thought of by many philosophers, as some kind of final, finished, timeless judgment of what's what, in this case of how best to classify something.)

If the assumption underlying the 'is–ought' gap were true, we could, as already noted, never derive anything from anything, since concepts and definitions are not established deductively but are the products of scientific and ordinary theorizing or formation of general principles. (This was seen, as I have noted, by Hume with reference to scientific knowledge as far as their having the status of deductive truth is concerned, but has later been accepted by such radical skeptics as Rorty and Feyerabend *vis-à-vis* any account of objective knowledge.) A concept is not formed by means of some intellectual intuition – a sudden, unexplainable 'seeing' – of a state of affairs (framed in a proposition) from which further states of affairs (or propositions stating them) may be formally deduced.

It is more sensible to understand concepts as formed by abstraction – by what Aristotle refers to as integration and differentiation, by thinking of what one is aware of and arranging this material in a coherent, complete, and practical order. We combine this concept-formation process with the development of language – which is, in large measure, an economization process, since it is easier to keep in mind and recall words than lengthy strings of ideas. And when we ask about the meaning of a concept, we do indeed ask on what sorts of occasions, with what conditions surrounding us, we will make use of it.[21]

The function of meaning is to carry out this process of concept formation objectively, and the nature of meaningful moral concepts – and, in the last analysis, a meaningful moral life – rests on such objectivity. It not only guides us to living rightly and well, but also enables us to know that we do so, should it become necessary to explain ourselves either to ourselves or to, for example, significant others.

As to the Moorean objection that any definition of what 'good' means can be earnestly questioned, let us note that such a question can be raised about definitions and theories in every branch of knowledge. When a new astrophysical entity, say 'dark matter' or a 'black hole,' is defined, questions abound and subside only when the definition has caught on. In ethics, politics, and aesthetics, however, such questioning rarely subsides since there are many motives for questioning – including attempts at obfuscation, professional devil's advocacy, and the need of the wisest members of each new generation to assure themselves of the truth of such personally significant questions.

Is and Ought in Nature

It is worth noting that outside of ethics, inferences from the 'is' to the 'ought' are common and unproblematic. I have in mind such life sciences as botany, ecology, biology, zoology, and medicine. In these areas we confidently refer to good and bad states, conditions, attributes, prospects, organs, and so on, with no suggestion that these are good 'to us' alone. When a botanist judges a plant to be doing badly or well, his standard is the plant's flourishing. This norm is an objective feature of the plant, not an expression of subjective human preferences or tastes. A redwood tree can be blighted or a liver cancerous quite apart from any human opinion. Even if a patient were to prefer to die, this would not change the diagnosis of the disease. The health of living things is not subjective, but objective, determined by the objective conditions of its flourishing.

Human norms do, of course, differ from such natural norms. First and foremost, unlike other living things, our self-actualization requires more than favorable environmental conditions. It requires our own active choices. A flower has no choice about whether or not it will flourish. If it finds the right conditions, it blossoms as a matter of course. This is not the case with human beings. Individuals who have the best possible environments often turn out rotten because of the choices they make, just as other individuals rise above inhospitable environments by making wise decisions. Because so much depends upon our decisions, we need to make wise ones. This is the purpose of moral philosophy. This is why human beings need philosophy and rosebushes do not.

Second, for human beings, individuality is much more important than in other living things. From an individualist perspective, nature is a depressing spectacle. Hundreds of baby sea-turtles hatch out and rush toward the water while predators swoop down to feast upon them. But if one in a hundred reaches the water, the species survives and nature (in the guise of the 'selfish' gene) is content. Millions of sperm cells fight for the privilege of fertilizing one egg. One sperm makes it, and the rest are expendable, yet nature is content. Human beings, however, are self-conscious. We are conscious of our individuality. Our individuality *matters* to us. It matters that we realize our individual natures, not merely expend them selflessly for the propagation of the species. And we value one another, when we do, as human individuals, not simply as humans or as indeterminate unique beings.

In spite of these differences, however, non-ethical norms throw valuable light on ethical ones. Doctors face no is–ought problem when they inductively infer what the healthy function of an organ is and then, when faced with an unhealthy specimen, prescribe medications to restore it to health. So too with moral philosophers, who inductively infer what virtues characterize a healthy human soul and then, when faced with a sick one, prescribe actions to guide or restore it to health. This is the basis of Plato's claim that virtue is the soul's health and the moral philosopher the soul's doctor.

As to the objectivity of moral claims, it may be objected to the above that this isn't the sort of objectivity that one finds *vis-à-vis* intrinsic properties, such as an apple's roundness or the wetness of water; it is objectivity nonetheless, of the sort we find in engineering or medical principles. So, as Jason Raibley notes about this conception of objectivity,

> though radically unlike Moore's, [it] shows us that values have all the 'fixity' that one could want. That is: [this] account is in any case a form of moral realism. We can make true evaluative judgments concerning the actions and ends of others. We can do this because most people have chosen life, and it is a normal, natural, scientific fact whether some subordinate action or end will be conducive to life. The right and the good are therefore both determinate and, within a certain range, universal. But … values are only objective insofar as they are the products of correct conceptual functioning, and [this] theory of values is objective only insofar as it is the result of objective thinking and congruent with a correct model of conceptual functioning.[22]

It is no fault of an account of objectivity that it shows that some kinds of claims can only be objective in one way, not in one that others are also objective. That is because the nature of the claims differs, thus making their establishment as true different as well. But they are true, nonetheless.

And in the case of value claims that bear on human conduct, there is that all-important matter as well, namely, that what is morally or ethically right for one individual to do may not be for another and yet another by virtue of who these individuals are, which is a central element of their human nature.

Notes

* This chapter is a revised version of Chapter 4 of Tibor R. Machan, *Classical Individualism* (London: Routledge, 1998).

1 The sense of 'objective' I apply to ethics and politics (as well as other field of human cognitive concern) derives from Ayn Rand's work, 'The Objectivist Ethics,' in her *The Virtue of Selfishness: A New Concept of Egoism* (New York: New American Library, 1961), and her *Introduction to Objectivist Epistemology* 2nd edition, ed. Harry Binswanger and Leonard Peikoff (New York: Meridian, 1990). This idea is discussed very fruitfully in J. R. Raibley, 'Rand on the Objectivity of Values,' presented at the 2002 Advance Seminar in Objectivist Studies, at UCLA.

It is vital not to confuse this sense of 'objective' with one that suggests that values are intrinsic, that they subsist in something (human life, nature, a painting). Objective values are, instead, relational, bearing on the value some conduct or state of affairs has for a being to which things can be of value. As Rand defines the idea, '*objective*, i.e., as neither revealed nor invented, but as produced by man's consciousness in accordance with the facts of reality, as mental integrations of factual data computed by man – as the products of a cognitive method of classification whose processes must be performed by man, but whose content is dictated by reality', *Introduction to Objectivist Epistemology*, p. 54. See, also, Allan Gotthelf, 'Ayn Rand on Concepts, Definitions, and Essences,' and James G. Lennox, 'Ayn Rand on Concepts, Context, and the Advance of Science,' papers presented at the meeting of the Ayn Rand Society, American Philosophical Association, 28 December, 2003, Washington, DC.

2 As per the dictionary: '1. a. Proceeding from or taking place in a person's mind rather than the external world: a subjective decision. b. Particular to a given person; personal: subjective experience … 3. Existing only in the mind....' http://www.yourdictionary.com/ahd/s/s0841800.html.

3 Norms or standards of conduct are problematic because they are supposed to pertain to the achievement of values or goodness, something that may be a property or condition of something. If the property or condition is objective, then the norm or standard can be shown to assess to what extent this property or condition is enhanced by some state of affairs or some action. If one acts because of and in line with a standard that realizes a value or good, this act is deemed to be right. The claim that such an act is right will then be true, that it is wrong, false. But if the property or condition is subjective or arbitrary – for example, it obtains at the arbitrary preference of someone who makes the claim – then the claim that such an act is right is not objective and cannot be shown to be true or false. It will merely be an expression of what one prefers or likes, desires or wishes. For more, see Tibor R. Machan, 'Rand and Choice,' forthcoming in *The Journal of Ayn Rand Studies*.

4 This book defends objectivity from some opponents in all spheres of *bona fide* knowledge but doesn't apply its possibility to all areas that are of deep concern to us. This chapter, in particular, focuses on an area that is of especially widespread concern among human beings, namely, whether when one holds that someone ought or ought not to do this or that, could one's belief or conviction be objectively true.

5 Milton Friedman, *Capitalism and Freedom* (Chicago, IL: University of Chicago Press, 1962), p. 12. See also Don Bellante, 'Subjective Value Theory and Government Intervention in the Labor Market,' *Austrian Economics Newsletter* (Spring/Summer 1989): 1–2.

6 Richard Rorty, 'The Seer of Prague,' *The New Republic*, 1 July 1991: 37. Rorty means by 'we' 'non-metaphysicians,' that is, those who understand that words and ideas do not represent some reality 'out there.' (There is more on this in Chapter 1 of this work.)

7 Richard Tuck, *Hobbes* (London: Oxford University Press, 1989), p. 116.

8 Even those who object to the claim that norms are objective make purportedly objective claims when they say that others, too, ought to reject this claim. The ought may be only a very mild moral rebuke in this instance, yet judging by the intensity and seriousness with which it is advanced, it is hardly to be taken as an expression of a mere preference.

9 It might be objected here that what I have listed are at most 'good policies' and that a policy cannot be true. Yet arguably the claim that these are good policies could be either true or false, which is just what the objectivist would try to show.

10 Those who deny the objectivity of moral norms, standards, or principles usually contrast this with their view that other claims pertaining to what science discovers or what we observe in the world around us are capable of being true or false, that what is being asserted in these areas, in contrast to ethics, politics, or aesthetics, is knowable, cognitively significant. Yet in these areas there is rarely universal agreement with the truths that can be ascertained, nor do all those who address the issues involved make identical knowledge claims. But those who uphold this 'is–ought' gap thesis would tend to account for the lack of universal agreement in terms not of the impossibility of reaching it but, rather, certain impediments some people face as they attempt to figure out what is true or to come to know what is the case. Of course, there are those, including Rorty, who would extend their skepticism not just to value judgments or moral claims but also to ones advanced in the sciences or other non-normative areas. They would hold that even in these the possibility of objective truth is an illusion. See, for more on this issue, Chapter 1 of the present work.

11 For a very well-worked-out application of this model, see Martha Nussbaum, *The Therapy of Desire* (Princeton, NJ: Princeton University Press, 1994) and Philippa Foot, *Natural Goodness* (Oxford: Clarendon Press, 2001). See, for a much earlier statement of a position along such (neo-Aristotelian) lines, Rand, 'The Objectivist Ethics.'

12 This point is very important in the debate about certain environmental issues. Some in that debate claim that nature is of value, intrinsically. See, for a different view, Tibor R. Machan, *Putting Humans First* (Lanham, MD: Rowman & Littlefield, 2004).

13 Raimond Gaita, 'The Personal in Ethics,' in *Wittgenstein: Attention to Particulars* (New York: St. Martin's Press, 1989), p. 127.

14 See for this example Jean-Paul Sartre, *Existentialism and Humanism*, translation and introduction by Philip Mairet (Brooklyn, NY: Haskell House, 1977).

15 Some utilitarians, such as Henry Sidgwick and, more recently, Fred Feldman, recognize this.

16 There is much here that could use expansion but I have addressed the issues in considerable detail elsewhere. See Tibor R. Machan, *Human Rights and Human Liberties* (Chicago: Nelson-Hall, 1975); *Individuals and Their Rights* (La Salle, IL: Open Court Publishing Co., Inc., 1989), and *Classical Individualism* (London: Routledge, 1998).

17 What is true is that, in the case of one who has abandoned life and is choosing to proceed on a more or less rapid descent to annihilation, no principled (moral) conduct is relevant. Morality is moot in death.

18 True claims of this sort are often probabilistic. There is always chance to contend with, and that is by definition unforeseeable.

19 David Hume, *A Treatise of Human Nature* (Garden City, NY: Doubleday & Company, Inc., 1961), p. 423. Hume argues that it 'seems altogether inconceivable, how this new relation [ought] can be a *deduction* from others' (my emphasis).

20 G. E. Moore, *Principia Ethica* (Cambridge, UK: Cambridge University Press, 1927). I want to thank Jason Raibley for his specific help with this discussion.

21 I draw here on the work of Ayn Rand, *Introduction to Objectivist Epistemology*, 2nd edn (New York: New American Library, 1990), as well as on Ludwig Wittgenstein, *On Certainty* (Oxford: Blackwell, 1969), and J. L. Austin, *Philosophical Papers* (Oxford: Clarendon Press, 1961).

22 Raibley, 'Rand on the Objectivity of Values.'

Chapter 6

Journalistic Objectivity, True and False

What are the moral obligations of journalists, publishers, scholars, and other conveyors of information? At the very least, conveyors of information are obligated actually to convey information; that is, they are obligated to tell the truth, to be accurate and objective, to be complete and fair. False, inaccurate, slanted, and incomplete information is not information at all. It is disinformation. Therefore, it is a disservice to the reading, listening, or viewing public and an evasion of professional responsibility.[1]

Independence, Neutrality, and Objectivity

Objectivity, however, is difficult. Indeed, a growing chorus of scholars and journalists claims that it is too difficult a standard for them to be expected to live up to. Merely because one is human, they claim, one will fail at being fully objective. Thus they beg the public's pardon to misinform them. Consider this comment from Anna Quindlen:

> The problem is that the press sometimes seems to suggest that reporting is objective science, that there is no scrim between the reader and the information. But there is always a scrim. The scrim is the reporter. And some reporters manage to shave it to a shadow, while others get in the reader's way, not usually because of overt bias but because of the limitation of their talents. We always carry with us what we've learned and those we like. But the best reporters, whether among the old guys in hats who once populated city rooms or the lacquered consonant-cracking princesses of local TV news, use that only as a starting point. Objectivity is a goal. Curiosity is the way to get there.[2]

Quindlen's point rests, in part, on an idealistic conception of objectivity as complete impartiality and transparency. In line with that view, to be objective is always going to remain an aspiration, like a perfect circle or a frictionless plane. Yet, as the cliché has it, the perfect is at least as often the enemy of the good. In this instance, too, taking objectivity to be some kind of unattainable ideal has made it possible to excuse nearly all biased reporting.

One notion that is often confused with objectivity is journalistic 'independence.' Some argue that a good, conscientious journalist must remain independent. Yet, consider what it is such a person might be independent of. Is it values? Surely this is impossible – everyone acts with guiding values. The only difference is whether they face up to this or keep their values hidden from themselves. (Others surely can spot them in a jiffy!) Or independent of the editor? Why should this be a virtue? Editors are supposed to give direction to what they edit. They should hire folks who express values they believe need to be expressed. If they hire people helter-skelter, then so much the worse for their editorial competence. Or are they supposed to be independent of owners? Well, why, again? Why would people own a publishing outfit other than to make money while doing some proselytizing? Owners take the risk with the business, so they surely have the right and responsibility to make certain it does something worthwhile. And even when they lie back and do not do hands-on guidance, they cannot help but be on the minds of writers. If reporters don't like the direction, they ought to quit or argue about it.

Worse is when one believes independence is desirable, even though, of course, it is not. You get spectacles like NBC-TV's Tom Brokaw who just finished reporting on some big government corruption and then proceeded, after the commercial, to lament the absence of federal regulation of aquarium shows.[3] He probably had no idea that he was promoting a political agenda.[4] That is what happens when you fancy yourself to be independent – you become independent of consistency and good judgment. You report and express values helter-skelter, you don't become independent of them. It would be best if journalists just admitted to having certain values, kept them straight and allowed the readers to contend with them. Frank partisanship is far more professionally honest than feigned independence. Partisanship need not at all conflict with objectivity – with getting it right about what's important and what follows from that. One can be partisan in support of a sound set of values, akin to how a medical doctor or researcher is partisan in support of good health. Journalists, however, need not take on the burden of defending their partisanship, only of disclosing it, to be candid about their guiding ideas and ideals.

There is no value-free journalism. It is a myth – or a ruse. As a myth, it has a reputable pedigree: positivism, the view that facts can be dealt with rationally but values cannot. That idea is itself derived from the view that only what one can observe by the senses can be a fact.

Both of these ideas are full of difficulties. For one, the ideas themselves cannot be confirmed by observation alone – quite the opposite. Even in physics there are facts that aren't observed but inferred from theories. And as to values, they are very possibly a kind of fact, namely, about what is good and bad. We know a lot of what is good and bad – just ask all the environmentalists and supporters of various causes. They at least claim to know. Is it a fact that racism is bad? Or that prejudice is? Do journalists really wish to argue that that's just unsupportable opinion? But if those can be facts, why couldn't it be a fact that some stories should be covered more vigorously, others less so? And that some people or activities are more important to cover in a community than others? Of course these could all be facts, if one just bothered to think about it carefully and long enough. It isn't simple but that isn't decisive.

So independent journalism is indeed a myth. It may also be a ruse. A lot of journalists like to disguise their values and when it is claimed they have any, they just deny that anyone can be objective because, well, we all have prejudices. But they shouldn't play a role in our reporting. It lets one be lazy.

Instead, responsible journalism requires figuring out what is important and covering it well. That, in turn, is a matter of professional ethics and open to scrutiny.

Furthermore, there is no requirement in the profession of journalism or scholarship – as indeed in being a juror – that one be neutral, only that they go about their work without sacrificing objectivity. This is a crucial point, one that has been noted throughout the present work: it is one thing to demand objectivity, quite another to demand neutrality or non-partisanship. In the former case, an unbiased assessment of facts and values suffices. The latter, however, is frankly impossible. No one embarking on any purposive human activity is neutral or non-partisan. It is a myth.

But the myth has powerful intellectual pedigree. It derives from the belief, propounded most forcefully by logical positivists, that values are by definition nothing more than biases. So even a medical doctor who is interested in saving the life of patients is biased about the value of life, because that value is incapable of being shown to be objective. All values are simply preferences, none better founded than others.

The value-free stance of so many social sciences rests in large part on this legacy of positivism. And from it has emerged a somewhat looser, albeit paradoxical, idea about the alleged merits of neutrality, non-partisanship, or independence, in particular in journalism and scholarship.

The paradox is obvious: while values are impossible to establish, it is nevertheless believed very firmly that it is valuable not to mix values with facts! Surely if it isn't possible to show the value of anything, neither is it possible to show the value of being value-free!

The value-free stance is impossible and, when attempted, leads to duplicity. Pretending to be value-free when one cannot be is itself the violation of a well-established value, namely, integrity. As we saw in a previous chapter, the case for value-subjectivism is unproven and there is every reason to hold that some values are true, others are not, or, putting it differently, some things are of value and others are not. To whom, when, and so forth is another matter, of course, but the basic point is to learn about the values that apply.

Yet there is a much saner version of objectivity that one could embrace and also practice quite conscientiously, namely, taking full stock of what is important to report about some issue *within the confines of space and time*. Of course, what will give pause here above all is the reference to what is important – as if that is always going to be a purely subjective matter, thus contaminating this idea of objectivity from the start. But that is to fall prey to a philosophically flawed idea of what it is for something to be important. If we understand that it is human life that is the standard of what is important, this problem vanishes.

Indeed, most journalists are unencumbered by the mistake other than when they start talking about the issue. We all realize, don't we, that reporting on an invasion of ants in some neighbor's kitchen isn't as important as reporting on an earthquake or rise in unemployment or corruption in the system of criminal law. Once one appreciates the implicit value standard exhibited in such common-sense ranking of topics, one can get a clear enough idea of the fact that what is important is not at all subjective. At times, of course, it can appear to be so, namely, when journalists bend to fads or trends that are themselves unrelated to anything objectively important but merely reflect some whim of the public or of their colleagues.

Commerce and Objectivity

The profit motive is often derided as an enemy of objectivity. In journalism, this problem has two sources: the owners of media organizations and the advertisers who are the principal sources of revenue. Journalists feel pressured not to report objectively on matters that affect the interests of their employers and advertisers. Here again objectivity is confused with impartiality or neutrality.

This lack of 'objectivity' is most obvious in trade publications, such as the education profession's *Chronicles of Higher Education* and IBM's *Think* magazine, which exist merely to advance the interests of various industries and professions. Such publications are so obviously partisan that they fall into the categories of advertising, promotional literature, and lobbying, not journalism. They are, however, so involved in special-interest pleading that few people are fooled. Hardly anyone expects them to engage in 'objective' reporting and adhere to standards of journalistic ethics. Such publications are perfectly proper in their own right, so long as they meet whatever ethical obligations are associated with advertising or advocacy. And while not neutral, they can, within their context, be objective enough, as when they report on some aspects of the firm or industry they serve.

The same is true of government publications, even those of the United States Information Agency, which is supposedly dedicated to informing the deceived foreign populations of the truth about America, but is hardly a paragon of 'objectivity.' Here, too, something akin to propaganda or at least advocacy comes into play and thus any breach of journalistic ethics is a moot issue. But, again, it is not actually objectivity that is missing, but non-partisanship.

Similarly, certain religious colleges, policy institutes, and think tanks posture as educational or scholarly institutions but are, in fact, partisan advocacy organizations.

But *bona fide* journalists do often enough encounter conflicts between the obligation of objectivity – in both its valid and distorted sense – and the pressures of commerce. The Western-style free press is not free to do just anything if it wishes to stay solvent. Even as eminent and secure a publication as *The Wall Street Journal* would have trouble directly attacking Dow Jones, never mind if its editors held it to be journalistically justified. *The Journal* is even required to print all of the parent company's news releases verbatim.

It should be noted, however, that this problem does not constitute a critique of objectivity *as such*. The very idea that owners and advertisers can compromise objectivity assumes that without such influence objectivity is possible. If the possibility of objectivity is rejected, why be concerned with undue influences? If distortion is inherent in the human situation as such, journalists, scientists, scholars, and the like will all be unable to achieve objectivity, regardless of what causes the distortion.

It is also important to note that it is not just the profit motive that can undermine objectivity. Non-profit and state-funded information sources also face conflicts of interest. Many media organizations receive support from powerful, well-established, and influential organizations, including the Ford, Rockefeller, Earhart, Olin, and Carnegie foundations, all of which have their own agendas.[5] Some of these sponsors are actual government bodies, such as the Navy and the National Endowment for the Humanities. How can such media sources be expected to report objectively and offer criticism on matters relating to their funders?

Innumerable scholarly papers devote their first footnote to thanking the Lilly Fund, the Rand Corporation, the National Science Foundation, National Endowment for the Humanities, or some other private or government granting body for funds that made possible the research leading to the publication of the piece. In each of these instances it could be problematic if the scholar were to focus his or her attention upon the source of the support.

In spite of this fact, it is usually the objectivity of research sponsored by industries – meat, tobacco, pharmaceuticals, and the like – that comes under suspicion from the press, not research conducted with funding from the government or non-profit foundations. Yet even academic journals are not exempt from conflicts of interest. For example, social work journals tend more readily to accept papers for publication that report the positive results of social work than papers that report the negative results of such work.

Who is likely to be more vigilant in aspiring to objectivity: someone who works in the competitive market of journalism or someone wedded to the near state-oligopoly of university education? When some professors bravely decline invitations from private corporations that have a political agenda, why are they willing to keep their employment with state universities, which are clearly part of a system with a firm agenda of its own, a system insulated from meaningful competition that might challenge this agenda? Is objectivity only compromised by business interests, but not by interests of state?

Thus it is naive to believe that only the profit motive produces bias. Any form of funding, public or private, profit or non-profit, has the potential to conflict with objectivity. Since it is not possible to eliminate funding altogether, it is necessary to determine which form of funding is least likely to undermine objectivity. It is also necessary to look for ways of strengthening the obligation to objectivity against temptations to the contrary.

Since all forms of funding can undermine objectivity, the best guardian of objectivity is the existence of a free, competitive market in information – not because such a system guarantees that individual journalists, scholars, or publications will be objective and fair, but because competition ensures variety of different media outlets and sources of funding. Therefore, if one media outlet fails to report honestly on matters that affect their owners and advertisers, there is the possibility that another outlet with different owners and advertisers will report the story objectively. The main danger to the consumer is state-protected monopoly or oligopoly, and as anyone not wholly ignorant of economics knows, those are more likely within an economy that enjoys extensive government intervention than in competitive or free markets.[6]

Ethics, Professional and Personal

One may, however, object that the consumer may be protected by the free market, but this does not excuse a professional who succumbs to pressures that militate against objectivity. If the patron of a scholarly or public policy foundation does something of interest to the scholars or researchers, should not the organization study and report on this? Or if a major advertiser of a newspaper threatens to withdraw in case a certain story is aired, what should the reporter or editor do? This is not a simple matter at all, since competing values are at issue, not just doing the right or the wrong thing.

The first general point is that while professional ethics is every bit the objective (but more narrowly focused) code of conduct that one should expect from a sound ethical system, it is neither a complete nor a fundamental moral code. Beyond the principles of professional ethics are the basic ethical principles which everyone should invoke in life. In short, a principle of professional ethics is not anything on the order of an absolute moral principle or, to use Immanuel Kant's famous phrase, a 'categorical imperative.' It is, for instance, contingent on the impact that adherence to other moral principles will require of a person.

Thus while professionals in business ought, as Milton Friedman has pointed out, pursue the success of their enterprise, there can be exceptions to this – for example, in family or national emergencies. That is, in business ethics, it is a moral responsibility of an executive to further the economic well-being of the firm. But since this executive is also a citizen, a friend, a parent, or a spouse, surely other moral responsibilities will also guide his or her conduct?

Total, blind devotion to work is morally irresponsible, since citizenship and parenthood, too, require attention. And while no conflict need normally arise between these roles, in some cases that is not avoidable and a choice may need to be made. In emergencies one must use discretion and maybe put in abeyance what is ordinarily required in these other roles in one's life. Normally, however, the coexistence of worthy purposes can be expected to prevail and no sacrifice of the objectives of one role need be made so as to serve others.

Now the same applies in journalism or scholarship. One may find an interesting story about some major supporter, but because it would be suicidal to run it, one should choose some other topic. If the story is vital enough, it may be wise to point this out to some competitor. Total silence could be immoral, especially if the story itself involves immorality – say the funder is betraying some innocent party or some important value, perhaps violating some ethical precept or breaking a just law. And there is a point at which one may have to forgo the benefits of the support of one's organization or professional activity – for example, if the funder is stealing the funds it contributes or is making demands on the journalist or scholar that would require betraying professional ethics. Of course, as with most ethical problems, the difficulties one faces in a conflict of interest situation can only be hinted at in abstract discussions. They must ultimately be dealt with in concrete situations and they may be distorted if dealt with sketchily. One needs to consider the details.

In professions such as scholarship, journalism, writing, and so forth, the ethics of the profession derive from more basic principles. In other words, the principles involved in professional ethics are not the very general ones that a sound ethical theory will propose for virtually every situation faced by human beings. Rather these are hypothetical principles: if you face these situations, then act in such and such a way. Principles of journalistic ethics are not similar to the prohibition against murder or theft but more like edicts about therapist–client or doctor–patient relations.

In too many cases, outside the most basic and simple ones, the right thing to do will have to be determined from a very rich context of background information. But it is clear that it is false to charge that because there is the potential of conflict of interest there is something inherently problematic in the practice of the profession itself. Objectivity need by no means be jeopardized when one works for a company or organization, even if there can be a temptation to compromise that requirement. It is reckless cynicism to charge that whenever economic or other interests are involved in some undertaking (that is, in virtually any aspect of human life), one cannot act decently.

Specifically, conflicts of interest may be handled in a variety of ways, including simply shelving them or passing on the problem to those who can discuss them more effectively, with less risk of severe costs. No threat to objectivity arises from taking such a course. The idea that martyrdom is noble belongs to a very dubious ethical outlook, one that certainly was not developed with an eye to making one a success at living a human life. Rather the point was to attain success in another world. It would be wrong, from a sane ethical viewpoint, to seek actions that lead to the destruction of one's values and projects, unless there is something very fundamental at stake, in which case one is still preserving what counts the most for oneself, one's integrity.

Journalistic and scholarly ethics do not require suicidal policies. Courage, of course, requires taking some risks, weighing values, and choosing sides when conflicts occur. But courage is not the only virtue – there is prudence, moderation, thrift, honesty, and the rest. Each must be attended to and being reasonable in this task is the ultimate ethical responsibility of every human being, one that is prior to the more specialized ethical responsibilities related to one's profession, including journalism.

The central element of objective journalism is to recognize that journalism, as other professions, aims at some goal, presumably a goal that is itself demonstrably valuable. Roughly, this goal for journalists is to report what has been going on in a given region they have committed themselves to cover, with a clear idea of what is very important, what is not so important, and what is rather trivial, and carry out the reporting accurately and without undisclosed partisanship.

In a society with a free press, this task is not going to be carried out flawlessly but, because of competition, consumers of what the media report will have a greater likelihood of obtaining full and objective coverage than in one wherein journalism is under government control or owned by a protected monopolist (or a few protected oligopolists).

Notes

1 Although these appear to be obscure disciplines, essentially they apply common-sense ethical precepts to special professional activities. For a discussion of business ethics, for example, see Tibor R. Machan, 'Ethics and its Uses,' in Tibor R. Machan (ed.), *Commerce and Morality* (Lanham, MD: Rowman & Littlefield, 1988). For an introductory discussion of ethics in general, see Tibor R. Machan, *A Primer on Ethics* (Normal, OK: University of Oklahoma Press, 1997).

2 Anna Quindlen, 'Journalism 101: Human Nature, The people of the press are people, too. So where does that leave the idea of objectivity?' *Newsweek*, 15 November 1999: 104.

3 There are those, such as Steve Kelman, *Regulating America, Regulating Sweden: A Comparative Study of Occupational Safety and Health Policy* (Cambridge, MA: MIT Press, 1981), who perceive government regulation as something almost apolitical, as a kind of non-partisan state conduct and when journalists share this view, their call for such regulation is probably taken by them to amount to no more than a call for food for a starving human being.

4 In the late 1960s and early 1970s a school of journalism became respectable and was dubbed 'activist.' It was associated most closely with investigative reporting, whereby journalists would embark upon the discovery of various facts about persons and/or establishments (firms, businesses, and other organizations) in order to identify them as

participants in the pursuit of goals the journalists deemed to be misguided, even perhaps criminal. See, for example, Joe Grimm, 'Combining Journalism and Activism,' in *The Detroit Free Press* (http://www.freep.com/jobspage/academy/schneider.htm). By the tenets of this school, a journalist would not breach professional ethics even if the work done advanced a particular cause (for example environmentalism, animal rights, free trade).

By implication, journalists who called for government regulation could be classified as activists, except that in the case of most of them, including Tom Brokaw, they would not identify themselves as activist journalists. See, for some discussion of this, Bernard Goldberg, *Bias, A CBS Insider Exposes How the Media Distort the News* (New York: Perennial, 2002). See, also, Elizabeth Wissner-Gross, *Unbiased Editing in a Diverse Society* (Ames: Iowa State University Press, 1999) and Hillier Krieghbaum, *Pressures on the Press* (New York: Thomas Y. Crowell, 1973).

5 There's a moral distinction between obtaining private versus government funding. In the former the support is voluntary, not expropriated, while in the latter the opposite is the case. This makes seeking government funding morally problematic, since arguably it amounts to inciting the state's violation of individual rights and unfairness (which violates the spirit of the US Constitution's 14th Amendment, whereby all citizens are supposed to be treated as equal under the law and, thus, none may enjoy special benefits from state action).

6 See Yale Brozen, 'Is Government the Source of Monopoly?' in T. R. Machan (ed.), *The Libertarian Alternative* (Chicago: Nelson Hall Publishing Co., Inc., 1974). As the issue relates to the telecommunications industry, see, among other works, Edwin Diamond, Norman Sandler, and Milton Mueller, *Telecommunications In Crisis: The First Amendment, Technology, and Deregulation* (Washington, DC: Cato Institute, 1983); Owen M. Fiss, *Liberalism Divided: Freedom of Speech and the Many Uses of State Power* (Boulder, CO: Westview Press, 1996); and W. J. Waluchow (ed.), *Free Expression Essays in Law and Philosophy* (Oxford: Clarendon Press, 1994).

Chapter 7

Richard Posner's Pragmatic Jurisprudence

Posner's Progress

Judge Richard Posner of the Seventh District Circuit Court and the University of Chicago School of Law is the chief advocate of a relatively mild yet not unimportant postmodern skepticism in legal studies. In his lecture 'Pragmatism versus the Rule of Law,'[1] Posner abandons his earlier positivist utilitarianism and embraces Rorty's brand of pragmatism.[2]

Although the two positions seem poles apart, there is a logic to Posner's transition. Positivism is so committed to empiricism that it holds that the only things ultimately knowable are sensory impressions. Once we realize that these impressions have to be interpreted by the mind in order to yield knowledge, we have transcendental idealism. Once we realize that these interpretations can change from place to place and from time to time, we have historicism. In any given period of history, the impressions are interpreted in line with prevailing traditions. Finally, since the impressions no longer offer any foundation, they are discarded, with only the beliefs of the prevailing tradition left, which have no basis in anything, not even the sense impressions to which hard-headed positivists appeal.[3]

Posner does not regard pragmatism as an alternative philosophy but as an alternative *to* philosophy: 'Pragmatism ... really isn't a philosophy at all. It is an anti-philosophy. It is destructive, but I think it is helpfully destructive' (19). He adds, 'I think the lessons of philosophy that I have gleaned and that are set forth in the problems of jurisprudence are entirely negative. The lessons are that systems, systematic thinking about morals, about justice, and about interpretation do not lead anywhere' (39).

In practical terms this leads to exactly nothing, although Posner declares that we should conclude that judges should make decisions on the basis of their intuitions:

> I am suggesting that we can, because we do, have confident beliefs without reasoning to them from unimpeachable truths ... because I haven't suggested and don't mean to suggest that our strong moral intuitions are true. They are merely undislodgeable at the time, an undislodgeable part of our grounds for action, and that is good enough for me, because I don't think we can do better. (8)

In other words, Posner thinks judges can have no basis for what they decide when what is called for is some moral judgment, say about justice or prudence or fairness, other than their intuitions or strong sentiments or feelings. Anything else is myth. Thus natural law is a myth. The attempts to rest, for example, US constitutional law on 'a natural law jurisprudence … are not convincing, … not rigorous, … not logical, … not coherent, … semantic, … arbitrary, and so forth' (68). Thus the best alternative to pragmatic intuitionism is a natural law jurisprudence that does not fail in these respects, namely, one that is convincing, rigorous, logical, coherent, not semantic, not arbitrary, and so forth.

My limited aim here is to question Posner's justification for embracing pragmatism. I will begin with a look at the standards by which Posner insists we evaluate jurisprudential philosophies. Next I will criticize Posner's claims about intuitions. I will show that his radical pragmatism can offer no remotely satisfactory account of jurisprudence or, indeed, anything else. I will argue that Posner leaves us with no such thing as jurisprudence but merely a bunch of robed individuals wielding arbitrary power, something they would be no more entitled to do than any terrorist wielding the power for which they are often condemned by robed men and women.

I should note, before going further, that I agree with Posner on several specific issues – for example, his largely free-market position and opposition to heavy-handed government regulation, his public-choice attitude toward the nature of government regulation of commerce, his general support of civil liberties, and even his critique of communitarianism (which puts him at odds with Rorty's social philosophy).[4] But I disagree with his account of how these views ought to be given support both inside and mainly outside the law.

Posner's Fallacy

Posner does not show that all attempts to ground jurisprudence in natural law are unconvincing. He merely asserts it. Being unconvincing seems like a serious problem, and it would be if we knew the criteria for judging an argument convincing. To convince someone requires that (a) the case be advanced logically, with true premises and sound definition of relevant terms, (b) the person one is attempting to convince consider such logical and factually well-grounded arguments a means to arrive at a conviction, and (c) the person pay heed to the effort being made.

But Posner tells us up front that he is 'making a standard pragmatist's gambit,' namely, 'attempting to dethrone logic from an important place in moral judgments.' He is 'denying the priority of reason in human judgment.' If reason does not matter, he will accept only what he intuits. Thus the fact that Posner isn't convinced by natural-law jurisprudence is more a matter of his own disinclination to consider such arguments than of the failure of those arguments.

Posner might respond that he is only doing what Hume and others have done throughout history, namely, rejecting logic in the realm of values while retaining it for epistemological and scientific purposes. But logic itself is a normative discipline. It tells how we *ought* to reason. If logic has been dethroned for normative purposes, then one cannot defend the following imperative: 'When one considers scientific and other so-called value-free disciplines, one ought to think logically, but not when it comes to ethics or jurisprudence.' Since this is itself an ethical proposition, Posner cannot insist that we be logical here but abandon logic elsewhere.

Posner claims that natural-law arguments are not 'rigorous.' This is a strange complaint given Posner's complete dismissal of objective standards of cognition. Clearly, there are different levels of rigor, depending on the subject matter at issue. When we advance arguments in mathematics their rigor will be different from that of those advanced in biology or psychology. Of course, in some fields an artificial rigor is introduced by rendering the verbal arguments that constitute the substance of the discipline into mathematical formulae. But in these we need to consider the level of rigor of the premises. When an economist uses the term 'preference' or 'utility' or 'satisfaction,' the issue is not whether the symbolic–mathematical representation of such terms in an equation is rigorous, but whether the terms themselves represent reality rigorously or precisely. And such terms will succeed at being rigorous in different ways in different tasks; precision will be different in biology from what is in physics.

Yet this is not all that can be said about Posner's complaint. By embracing Rorty's pragmatism Posner is actually denying that any terms can represent anything in reality. In Rorty's words, 'the notion that the reality referred to by [for example] "quark" was "determinate" before the word "quark" came along' is wrong.[5] So if Posner accepts Rorty's pragmatism, he has no grounds for demanding rigor from any argument. Rigor assumes that some determinate reality is meant by a term and could be represented more or less correctly and precisely. All Rorty can say for rigor is that, for example, 'the word "quark" … is useful for coping with the environment.'

If Posner finds this view sound, he will have no basis for rejecting natural-law jurisprudence or any other approach to the law – anything goes, as Paul Feyerabend (another radical relativist) argues in his *Against Method*.[6] Not that Posner is fully aware of this result. He may well wish to retain some semblance of common-sense realism as far as science and other factual judgments are concerned.

But as with many radical pragmatists, no such prospect remains because of their drastic anti-representationalism. The community of scientists whom they trust to remain loyal to common sense is, as Feyerabend's work makes plain, far from cohesive, and unless we are able, epistemologically, to tie this community's procedures to certain commonly recognizable, independent, objective facts of reality, their authority as dependable witnesses to special truths will be unfounded and arbitrary.

Posner claims attempts to establish natural-law jurisprudence are 'not logical.' But there is logical and then there is logical. Can one establish a moral principle deductively? Hume showed that deductive arguments, based on simple empirical premises, do not support moral judgments, simply because the premises do not include the term 'ought,' whereas the conclusion does. Thus the argument must be invalid. But as Hume also demonstrated, on this line of reasoning none of the sciences are logical. No series of premises containing 'is' judgments can lead to a conclusion including a 'must' or 'will be.'[7] Thus efforts to establish scientific principles are 'not logical' either. Nor indeed are efforts at establishing any judgment outside of a purely stipulative formal system.

Posner is not aware of how deeply the skepticism of both Hume and Rorty cuts. Thus he claims that,

> In certain human activities, there is a reality check that enforces a high degree of consensus. A scientist proposes a hypothesis that can be tested by a controlled experiment. An engineer designs an airplane that either will fly or will not fly. In both cases, we have an external check against these people's claims. (17)

Hume and Rorty would disagree, but the most vehement critic of this position is Feyerabend. Rorty notes that Feyerabend, like himself, 'is really against the correspondence theory of truth.'[8]

But it is the crux of Posner's covertly assumed theory, namely, that truth consists in a relation between what is said and what there is, or that our true claims correspond to facts that can exist independent of our awareness. Posner might claim that even without such a realist account we can still recognize something like useful or workable or practical knowledge. But speaking along these pragmatist lines simply avoids the hard issues, namely, what exactly is knowledge, and why is it better than hunches or opinions? To show that a belief is useful, we will have to appeal to knowledge that is more than useful. Indeed, we cannot say that Newton was wrong at all, unless we have more than the usefulness of beliefs at our disposal, for then all we can say is that, perhaps, Newton wasn't as useful as Einstein seems to be now. Yet such temporal ways of speaking run aground quickly – it could be said that Newton was just as useful in his era as Einstein is in ours, whatever the truth of their respective theories. We need something better than a pragmatist account of knowledge to clear up the evident confusion.

Posner's point that natural-law arguments are 'not coherent' is redundant, since coherence is determined by means of logic. If these efforts are not logical, they could not be coherent either.

Yet 'logical' can also mean that an effort to prove something is consistent with the requirements of Occam's Razor and other broad principles of rationality. This is how the term 'reasonable' is used frequently in law. Accordingly, one could reason that nothing explains human behavior better than the assumption that people have a capacity for rationality, that they can think and that they can do so more or less well, more or less carefully, more or less uncompromisingly.[9] This is the best explanation for human behavior. Black holes and the intentions of criminal defendants are all established by means of logical reasoning, in the broad but not necessarily deductive sense of that term.

And if Posner thinks that natural-law jurists have failed to be rational in this sense, he is probably wrong. For natural law means norms of conduct that may be established by arguments to the best explanation, starting with the idea of human nature that makes the most comprehensive, most consistent – but not necessarily final and unrevisable – sense of human affairs.

Is the effort to establish natural-law jurisprudence in vain? Not if the concepts employed by natural-law theory have a realistic foundation. But it is just this foundation that Posner and Rorty deny when they accept anti-representationalism, saying that words are merely 'useful for coping with the environment.' But this characterization is hopelessly question-begging. How are we to understand 'useful' or 'coping' or 'with' or 'the environment' if terms never represent reality? What will establish whether some words do or do not help us cope with the environment if these terms represent nothing that can make sentences in which they are used true or false? And what does the charge of 'arbitrary' amount to on the lips of someone who embraces Rorty's viewpoint, according to which every judgment is without foundation and merely may or may not have the benefit of community support?

None of Posner's charges against natural-law jurisprudence carries any weight, even within Posner's pragmatist viewpoint. The terms belong to a frame of reference rejected by Posner. Nevertheless, this frame of reference is quite effective in making sense of our cognitive experience, from common sense through science all the way to natural-law jurisprudence.

Impotent Intuitions

Posner claims that since natural-law jurisprudence is a failure, we must settle for an intuitionist jurisprudence. This means that we have some 'undislodgeable … grounds for our action' which are, however, not true but simply beliefs in which we have confidence. Who exactly *we* are throughout these claims is a mystery, for clearly different people in different communities and cultures do not share these undislodgeable intuitions.

Even the same person finds his or her intuitions dislodged and replaced from time to time – sometimes even as a result of having been convinced by another that what he or she believes is wrong. Prejudices are often dislodged even when people have held them for some time as 'grounds for … action.' When I was a child, I had convictions about Jews that I later abandoned because subsequent experience showed the arguments supporting them to be ridiculous. If one lives long enough, one witnesses such dislodgings often enough.

Here, too, Posner might deploy his fact/value gambit by claiming that what I learned had to do with facts, not values. Yet if the norm 'one ought to respect facts rather than prejudices' were not a sound one, the factual issues would have no bearing on the matter. It does not matter if one's prejudices are not founded on facts unless it is true that we should not be guided by them.

Posner thinks there are other ways 'in which one can discuss these highly charged, emotional questions,' for example, the wrongness of Nazi ideology, than natural-law jurisprudence. He thinks, for example, 'that the Nazi ideology was founded on a number of empirical mistakes concerning race and other matters, which helped explain why Germany was defeated and which have helped discredit Nazi ideology ...' But he adds that he, 'would never try to [prove] that genocide is a moral wrong.'

But as a Rortyean pragmatist, Posner cannot even make sense of the idea of 'empirical mistakes.' Posner fails to appreciate just how deeply Rorty's pragmatism cuts against realism. Different communities will find different empirical claims to be acceptable. It is not just moral intuitions that cannot be verified, but judgments about empirical matters as well, if such judgments cannot represent an independent reality. An empirical mistake is, after all, a realist notion. So on Posner's own terms, the Nazis cannot be faulted for their immorality or for their factual mistakes, contrary to what he says. It seems clear, then, that Posner does not have any way to handle 'highly charged, emotional questions' about Nazism, communism, child molestation, or the practice of murdering wives to obtain new dowries in parts of India.

Perhaps, then, Posner should simply admit that he has not adequately explored the basic questions of moral philosophy and defer to others – just as he, wearing a judge's robe, would insist that others defer to him on matters of positive law. Instead, though, he simply embraces a totally empty skepticism. But this is as intellectually and morally irresponsible as giving medical advice without adequate study of medicine.

Arbitrary Moralism

Thomas J. Bieter takes a dim view of Posner's pragmatic jurisprudence:

> The 'value neutrality' of judge Posner reminds me of the jurists and lawyers in the Nazi era. Having rejected the natural law tradition's connection between fundamental morality and law, and opting for law as the will of the state in legal positivism, they were intellectually disarmed and thus neutral when the Nazis began their assault upon the Jews. And this public 'value neutrality' of the intellectual, whether in or out of the university, influences the many at the margin of society and again makes a David Duke a possibility.[10]

There is some dispute as to whether their positivist, value-neutral stance did disarm judges in Nazi Germany, as Bieter and others claim. But if reason is irrelevant to normative judgments, then nothing follows as to what we should or should not say or do, and anything goes, including Nazi-style politics. A person who is neutral about values will not judge some action morally good or bad, right or wrong, no matter how abhorrent it may be. All he will observe is that the conduct does not please some people. Thus, as we saw with Rorty, the chief pragmatic value of realism is that it allows a rational assessment of the merits of tyrannies versus democratic societies.

But what about Posner's intuitions? We can at least hope that he is committed to respecting the law of the land. But with Posner this would not be a principled stance, merely one to which he was committed back in July 1991. Posner's description of his intuitions as 'undislodgeable at the time' is a neat bit of rhetoric, but not very comforting. For what is a conviction that is undislodgeable at a given time but a conviction that can be dislodged at another time? Given that he completely rejects objective moral truths, it is not difficult to envision Posner changing his views if that works for him. This, incidentally, pretty much characterizes the conduct of amoral gang members and sociopaths. Because of his Rortyean anti-philosophy, he is likely to remain dogmatically skeptical about any efforts, past or present, to ground moral judgments, including the judgment that a judge should be loyal to his oath of office and follow the constitution.

Accordingly, we have no reason to count on Richard Posner to be a friend of the rule of law – quite the contrary. Law, in fact, cannot be more to him than a set of arbitrary edicts laid down without rhyme or reason, without logic or philosophical or moral grounding. Not even the will of the people need matter to Posner. Indeed, by his own account, there is no reason to trust him when he tells us anything, if we are to take his anti-philosophical declarations at face value. He may be pulling our leg, for all we can say.

But even more alarmingly, Posner's account of the nature of law makes it no more obligatory than the pronouncements of a terrorist or a raving lunatic, so long as they can create some sort of communal consensus. In the last analysis, then, by Posner's own account the difference between a judge of the Seventh Circuit Court of Appeals and a wild-eyed terrorist is merely a matter of garb, and even that difference can be looked at in any way one wishes – the terrorist could see his as a point of rebellion and Posner's as the arrogance of power – nothing else.

I think we can do far better than Posner's account of law. As Anthony Kenny put it very poignantly,

> Just because the criterion for correct belief given by the classical foundationalist fails, we should not conclude prematurely that no criterion can be given which will help to distinguish between rational and irrational beliefs, between sense and folly, between sanity and madness.[11]

I have not, however, tried in this particular work to develop a case for natural-law jurisprudence.

I have suggested, however, that Posner's objections to such a jurisprudence are weak and inconsistent and that the details of such weakness and failing are so deep-seated that they render Posner's Rortyean pragmatic jurisprudence nearly meaningless. We might note, simply in passing, that if Posner wishes to retain his confidence in empirical studies, scientific findings, and other factual investigations related to the cases before him, he will have to grant the possibility of sound moral notions as well, if only because moral notions are as much a part of the world as the others.

So however difficult it might be for a judge to find moral grounds for a decision – or, more appropriately, for the basic laws of the land on which decisions rest – the attempt must continue, because, to paraphrase Posner, 'I don't think we ought to do less.' If this is true, then Posner's pragmatic jurisprudence is really no jurisprudence at all.

This presentation, too, then, has been a destructive analysis but, to recall Posner's own arguably dubious characterization of what he set out to do in his discussion, 'helpfully destructive.'[12] It has been helpful, I believe, in showing that those who tilt against natural law tend, in the main, to undermine their ability to make any recommendation about how we ought to act, whether as judges or legal theorists, scientists or ordinary people. Rorty and Posner, like so many other radical skeptics, are unable even to doubt what they doubt, since they undermine even their capacity to engage in intellectual destruction, let alone any constructive work.

Let me note, finally, that even though Posner may not wish to embrace the radically anti-representational and anti-rational conclusions his premises imply, surely he bears some responsibility if his students draw those conclusions. For those accustomed to legal reasoning, setting precedents on the relationship between law and morality should not be treated in a cavalier manner. These are serious matters, affecting the lives of individuals and the courses of nations. It is regrettable that Posner does not deal with the foundational issues of jurisprudence with the same seriousness he applies to their concrete applications in the courtroom.[13]

Notes

1 Richard Posner, 'Pragmatism versus the Rule of Law,' American Enterprise Institute lecture, 1 July 1991. I shall cite the pages of the transcript parenthetically in the text. Since the writing of this chapter Posner has published *Law, Pragmatism and Democracy* (Cambridge, MA: Harvard University Press, 2003). It is interesting to note that although Posner shares many of Richard Rorty's views, he is positioned on the political Right, while Rorty aligns himself with the political Left.

I should also note here that Posner's pragmatic approach to jurisprudence has had a very respectable pedigree. Justice Oliver Wendell Holmes, Jr, for example, was an explicit champion of American pragmatism, although this position is very far from the radical pragmatism of Rorty. Contemporary pragmatists who would not align themselves with Rorty include Susan Haack and the late Sidney Hook, among others. (See Susan Haack, *Evidence and Inquiry: Towards Reconstruction in Epistemology* (Oxford; Cambridge, MA: Blackwell, 1993) and Sidney Hook, *The Quest for Being* (Buffalo, NY: Prometheus Books, 1991).)

2 It is not clear whether Posner recognizes just how drastic Rorty's skepticism is, because in portions of his discussion he seems to continue to embrace a form of positivism, according to which, although we cannot affirm moral or political claims to be true, we are able to do this with empirical claims. But Rorty's pragmatism makes all claims equally theory-dependent. Only from within a community with a given frame of reference can one affirm or deny some claim, there being no independent objective reality against which such claims, whether empirical or normative, may be checked.

3 Rorty, *Objectivity, Relativism, and Truth*, p. 5.

4 Richard A. Posner, 'Why Conscription Does Not Serve Community, An Army of the Willing,' *The New Republic*, 19 May 2003.

5 Rorty, *Objectivity, Relativism and Truth*, p. 5.

6 Paul Feyerabend, *Against Method* (London: New Left Books, 1975) and *Science in a Free Society* (London: New Left Books, 1978).

7 For a very illuminating discussion of the entire fact/value, description/evaluation dichotomy, and why it has been completely misconceived since Hume, see Julius Kovesi, *Moral Notions* (London: Routledge & Kegan Paul, 1968). It bears noting that what exactly Hume wanted to achieve with his distinction is not entirely clear. He may only have aimed at dissuading us from thinking that one can deduce 'ought' statements from those containing the 'is' copula. But Hume also believed that no deduction of 'must' or even 'will' statements can be made from those containing only the 'is' copula. The point may simply be that our reasoning does not consist of pure deductive inferences in any area of discourse.

8 Rorty, *Objectivity, Relativism, and Truth*, p. 28n.

9 Tibor R. Machan, 'Applied Ethics and Free Will,' *Journal of Applied Philosophy*, 10 (1993), 59–72.

10 Thomas J. Bieter, 'Letter to *TIKKUN*,' *Vera Lex*, 12 (1992): 43.

11 Anthony Kenny, *What is Faith?* (New York: Oxford University Press, 1992), p. 14.

12 For my own efforts to produce a convincing case for natural law, see Tibor R. Machan, *Individuals and Their Rights* (La Salle, IL: Open Court Publishing Company, 1989); 'Law, Justice and Natural Rights,' *Western Ontario Law Review*, 14 (1975): 119–30; 'Human Dignity and the Law,' *DePaul Law Review*, 26 (1977): 119–26; 'Metaphysics, Epistemology and Natural Law Theory,' *American Journal of Jurisprudence*, 31 (1986): 65–77; 'The Unavoidability of Natural Law and Rights,' *Modern Age* (Winter 1987): 38–44; and 'Is Natural Law Ethics Obsolete?' *Vera Lex*, 9 (1989). Furthermore, I discuss what I call minimalist foundationalism in Tibor R. Machan, 'Evidence of Necessary Existence,' *Objectivity*, 1 (Fall, 1992): 31–62. I argue that we do have axiomatic (which does not mean infallible, timeless, or unchanging) knowledge of some facts that function, in part – if we but choose to keep them in focus – to secure for us objectivity and realism in all our fields of inquiry, provided we also attend to the details we can become aware of concerning those fields. Indeed, this accounts for why even the most skeptical of skeptics, such as Feyerabend, resort to faulting people for their lack of consistency, why reasoning well is a normative requirement in any discourse, since the laws of logic identify the most general, universal, and basic facts of being, the denial of which immediately leads one astray. They also govern concept formation, supplemented with such rules as Occam's Razor, parsimony, and so on.

13 I wish to thank J. Roger Lee and Richard Posner for their criticism of an earlier version of this discussion.

Chapter 8

Terrorism and Objective Moral Principles

I

I wish to maintain in this last chapter something that arises out of a consideration of the nature of terrorism. I argue that terrorism presupposes certain basic and probably universal moral principles, binding upon human beings. It may seem that I am discussing this issue in light of current events, specifically the terrorist attacks of 11 September 2001, but that is wrong. I had written an early version of this chapter back in 1987 and have reworked it here because of its potential for shedding additional light on the topic of objectivity.

The essentials of the argument follow: terrorism involves as a crucial feature the threatening of innocent life and its sacrifice. As such, it poses a dilemma to those who aim to combat it or even protect against it. Terrorism cannot make sense unless the dilemma has moral significance, that is, unless all alternatives are in fact morally repugnant. A response to effective terrorism amounts to having to decide which is morally worse, violating the rights of innocent human beings or failing to stem future threats to, and sacrifices of, innocent people. The outstanding feature that terrorism exhibits from the point of view of decision-making is that the alternatives are almost uniformly morally repugnant. And this depends on the objective moral evil of certain courses of conduct, at least objective for normal circumstances.

Does terrorism involve, essentially, the threatening and sacrifice of innocent human life? That is not all, but in case after case, barring some confusion, terrorism and other crimes involving assault upon, or threats to, human beings are differentiated by reference to the deliberate utilization of innocent human life. Terrorists cannot achieve their goals, which they regard as righteous, without threatening that, if they are not accommodated, something morally repugnant will be the result. That usually involves the death of innocent human beings or the violation of the rights of innocent people at the hands of the very people who are sworn to defend innocent human beings.

'Objectivity' means that whatever is objective can be shown to be true and, in the case of moral principles, binding on human beings. Terrorists may or may not regard some principles objective and binding at all. For instance, the members of the Palestine Liberation Organization probably regard the principle that connects them to lands under Israel's present control (and, by the majority of its citizens, its justified jurisdiction) as objectively true and binding on all, including Jews, and as one that any sensitive person should regard in a similar light. Yet, quite aside from this, some actions of Palestinean, Al Qaeda, and other terrorists – for example, taking defenseless airline passengers hostage, blowing up planes with utterly innocent passengers such as students on class field trips and office workers – also testify to the belief that other people regard certain principles as objectively binding on everyone. Furthermore, terrorist acts as such can be regarded as unintended means of underlining the objective validity of the principles that the terrorists wish to get others to heed.

Terrorists are incomprehensible without the acceptance of the plain fact that they are morally wrong. Terrorism threatens and sacrifices innocent and defenseless persons and it exploits the fact that others would usually prefer not to be party to such conduct. Often reaction to terrorism leads to violation of the rights, sometimes including the murder, of such victims. In the process of attempting to avert terrorist successes, some authorities will disregard the rights of bystanders, even the victims of terrorism. But this is not a necessary consequence of successful reaction to terrorism. It is more appropriate to view some of the harm that comes to innocent bystanders or victims as induced by the terrorists, with the responding authorities only constituting the instruments of such harm. For example, if, in reaction to a terrorist attack which threatens the loss of lives, someone inadvertently shoots an innocent bystander and no safer way of making an effective response was accessible, the total blame for the harm to the innocent bystander must go to the terrorist. The general point here is that reaction to terrorism is not something initiated by those doing the reacting not at least the manner of the reaction, which is often physical force.

A second point is that, although a principle of morality may be objective, it need not be absolute or categorical. Moral principles may apply very restrictively; yet, it may be demonstrably true that in such circumstances where they apply they do indeed apply. Only the soundness of some moral principles – their objective validity – is presupposed in terrorism.

II

Terrorists place authorities in the position of having to decide between virtually equally unacceptable courses of conduct. For instance, should terrorists be let go free *or* should the civil liberties of bystanders and even victims be threatened? Opponents of terrorism are left facing moral dilemmas.

Of course, strictly speaking, there may be no moral dilemmas at all. In short, it may in fact be possible in all cases to resolve what appear to be problems based on the moral impossibility of choosing any of the existing alternatives that one faces. For a solution could consist of flipping a coin, in the last analysis. But this itself shows how desperate a situation a moral dilemma amounts to. If human beings, doing their very best to figure things out, cannot come up with a better reason for doing one thing rather than some other than the mere fact that chance directs them to it, especially when the consequence of doing the act could be awful indeed, then these human beings surely must be facing a very terrible situation. Since, in response to terrorist acts, it is usually not only several officials of some government, but sometimes officials of several governments and private organizations – for example, in Munich, during the 1974 Olympic Games, the German, Israeli, and other governments, as well as various private organizations, such as the International Olympic Committee – that have to decide what to do, terrorist actions are ever so difficult to confront. One individual might rely on some moral point of view which provides him with guidance in even the worst human circumstances. But when a dozen or so individuals must forge a course of reaction, often representing their government and having to live with the opinions of millions following their decision, the situation can become virtually impossible to handle.

All this renders the terrorists' actions rather effective. Yet it is just this effectiveness that suggests that terrorism assumes something that many of those who wish to understand it and to cope with it would probably deny. This is that an objective standard of right and wrong exists and is, at times, almost universally admitted – at least, it shows in how human beings conduct themselves, apart from whether they would or could give full and cogent expression to the standard. Terrorists assume that the bulk of humanity, especially the law enforcement officials of most human communities, regard the threatening and sacrifice of innocent human life to be morally wrong. This makes clear sense of why those who are called upon to react to terrorism face the moral dilemma of either threatening and sacrificing innocent human life or tolerating this on the part of others, namely, the terrorists.

III

It will now be useful to examine what practical implications might be drawn from the points made above for purposes of coping with terrorism. The major difficulty in terrorism has been precisely the issue of what to do in the face of the moral dilemma which terrorists create. I do not propose to solve that problem here. But I think I can clarify what can be learned from the simple fact, if it is a fact (as I briefly argued), that terrorism makes sense only if objective moral principles exist.

Put bluntly, people who are faced with terrorism need to accept as fact that they are right about some things, that they stand in support of certain objective moral principles, and that conscientious actions performed in behalf of these principles are themselves morally praiseworthy. In short, those who combat terrorism need to accept the fact that they are, with respect to combating terrorism, morally right. Let me amplify.

Terrorists are characteristically righteous and, as noted before, sincere. Terrorists are not merely hoodlums or pranksters or even gangsters. Terrorists see themselves as crusaders. As David Rapoport observes, terrorists aim to 'dramatize the cause, prove that someone is willing to risk everything for it.'[1] Rapoport quotes Nachaeyeff, 'the first Russian terrorist of the 19th century,'[2] as claiming that 'the aims of our Society are … the entire emancipation and happiness of the people [to be] brought about only by means of an all destroying popular revolt.'[3] They, like all who believe that the end justifies the means, must provoke government to 'employ all its power, all its resources towards increasing the people's calamities and evils until their patience is exhausted and they break out in a *levee-en-masse*.'[4]

Judging by the terrorist's sincere willingness to sacrifice himself and others for a cause unrelated to personal gain or whim, the expectation that those combating terrorism will lack the will to act is not irrational, especially in the context of dealing with officials of Western liberal societies. In these societies, the ideal of tolerance for diversity is widely embraced. This idea is based on the view that individuals are indeed different from one another and often what will be right for one will not be right for another. And this ideal has come to be interpreted by many as moral relativism. Indeed, many officials and persons speaking on behalf of liberal societies hold that no moral code is binding on all human beings,[5] that moral principles are relative at least to cultures, but even, very likely, to groups and individuals within cultures.[6] Is this not, they would ask, the reason for the doctrine of tolerance?[7] Is this not why the law refuses to align itself with any moral position and protects everyone against any attempt to impose some code on individuals that they do not wish to accept freely or uncoerced?

I have voiced some of the themes of liberal democracy, which are familiar enough. One can recall some of the contentions of neoclassical economists,[8] who think, on the whole, that everyone's tastes and preferences must be granted equal validity; or of existentialists who deny that anything like objective, universal values can be identified regarding human life;[9] or of moral skeptics who believe that from respect for facts – that is, a proper scientific attitude – nothing concerning what we should do or should not do (the famous 'is–ought' dichotomy) may be concluded;[10] and so forth.

In contrast, terrorist groups are not slowed down by the worries listed above. For example, of hundreds of renditions of Marxism, certainly one[11] provides a justification for precisely the sort of thing that Nachaeyeff put so bluntly. Terrorism is OK as long as 'the entire emancipation and happiness of the people' are at stake. The 'new man,' *à la* Marx, Lenin, Stalin, Mao, or some more recent but perhaps less strident adherents, must be brought about, come what may. The moral imperative is precisely to do this and not to be deterred by bourgeois values, such as individual rights to life, liberty, and property, for example.

It is wrong to think of liberal, Western democracies as lacking moral foundation. These regimes, more than all others, have an objective and universal moral basis. Furthermore, it is a mistake to think that tolerance implies acceptance of relativism.[12] In fact, tolerance of diverse value systems rests on the moral truth that every individual must come to learn of the objective values that are there to be identified as a matter of choice, that it is wrong to force someone to 'see' truth.[13]

The tolerance that is part and parcel of Western liberal democracies concerns the value of individual responsibility. It relates to respecting individuals as having the capacity to learn of moral truth and of their need to be willing to do this learning lest they fail to achieve moral stature. This surely has nothing at all to do with relativism and, ironically, terrorists testify to this in their acts of terrorism.

IV

I have suggested that terrorism makes no sense unless some moral principles are objective, even universally binding on human beings. I have not shown that such principles exist. But terrorism as a real phenomenon presupposes them. For instance, terrorism presupposes that causing the death of defenseless or innocent human beings is morally wrong. Those who are faced with having to cope with terrorism usually act in a way that testifies to the objective and binding character of this moral principle. And they recognize this, as they choose between equally objectionable alternatives – for example, of contributing to the death of defenseless or innocent human beings or of leaving terrorists free to embark on future terrorist acts.[14]

This may suggest that terrorism is impossible in attacks upon a morally corrupt legal system, for example in the Soviet Union or in Nazi Germany, and that is correct, in two senses. Attacks upon systems which do not involve defenseless and innocent people are not terrorist attacks, but are, rather, forms of political resistance or rebellion. Attacks on tyrannies utilizing terror and involving defenseless and innocent people cannot have the same effect that terrorism can have against reasonably decent regimes. A corrupt, tyrannical system is not morally committed to policies which make terrorism possible. A tyrant, if effective, will simply seek to eliminate the attacking force even if defenseless and innocent lives are destroyed in the process. (This is why terrorism was rare in, for example, the USSR.)

These points suggest one practical implication. Those faced with the need to cope with terrorism should accept that they stand for certain objective moral principles. In other words, those making policy concerning terrorism and those in the field who must implement such policy should realize that the existence of terrorism presupposes certain objective moral edicts, for example that contributing to the death of defenseless and innocent human beings is morally suspect. (This is just an instance of a possible objective moral edict and does not exhaust all the moral edicts that might have a bearing on the situation that terrorism creates.)

The very systems vulnerable to terrorism, namely, Western liberal democracies (or the various more or less faithful versions we find today), tend to eschew objective moral principles and are indeed exceptionally open-minded and tolerant concerning moral matters. Such reluctance and tolerance are widely misunderstood, even by those practicing them. There is a better way of characterizing Western liberal democracies – or their ideal versions. This is to acknowledge their adherence to some objective moral principles. Their systems are, furthermore, justified to the extent that these principles are being adequately defended by the governing regimes. Among those objective moral principles, we find that moral conduct will not be enforced unless the rights of others are being violated. The reason terrorists are to be fiercely fought is that they wish to implement their convictions by violating others' rights.

So conceiving the issue avoids the problem of having to reconcile the usual policy of tolerance with the policy of intolerance toward terrorists. The view that Western liberal democracies are value-free systems, that democracy is not compatible with the legislation of morality, that freedom and moral values are incompatible – this view has contributed to some of the difficulties the West has in dealing with terrorism. For example, the idea that harshness with terrorists is authoritarian – a form of imposing one's values on others without good reason – is stagnating in an effort to cope with such outright enemies of human life as terrorists clearly are.

Abandoning the false conception of the character of Western liberal democracies is important for a variety of reasons. In connection with dealing with terrorism it can help in the development of a more realistic conception of what one is and should be doing. For example, coping with terrorism should perhaps be seen as coping with unmitigated evil, not with just some different people who hold different opinions from one's own; as one devises the methods that may be employed in fighting terrorism, one must be clear about the reasons why that kind of fight must be waged. One should be clear, also, about what sort of trade-offs, in moral terms, are justified when that kind of fight has to be waged. It is crucial, furthermore, that the general public be aware that those who wage a fight against terrorism regard this fight justified and an expression of their responsibility of public service.[15]

Notes

1 David Rapoport, *Assassination and Terrorism* (Toronto: Canadian Broadcast Corporation, 1971), p. 47.
2 Ibid., p. 40.
3 Ibid., p. 47.
4 Ibid.

5 Christian Bay, *The Structure of Freedom* (Stanford, CA: Stanford University Press, 1958), p. 33: 'Standards of morality and justice are what Hume calls "artifacts"; they are neither divinely ordained, nor an integral part of original human nature, nor revealed by pure reason. They are an outcome of the practical experience of mankind, and the sole consideration in the slow test of time is the utility each moral rule can demonstrate toward promoting human welfare.' This passage is quoted approvingly by F. Hayek, *The Constitution of Liberty* (Chicago: University of Chicago Press, 1960), p. 436. Hayek is the foremost contemporary representative of classical liberalism, the main philosophical foundation of Western democratic liberalism. Elsewhere, Hayek makes the point somewhat differently, asking rhetorically, 'Well, isn't the idea that society or the political area is bound to ethics, a view of what is right for man, almost incompatible with the idea that a man ought to have freedom to decide on what he regards as right?' *Reason,* February 1974, p. 11.

6 R. Beardsmore, *Moral Reasoning* (New York: Schocken, 1969). Beardsmore presents here a prominent view of morality, associated with Ludwig Wittgenstein, to the effect that moral matters are to be considered within a given moral system – for example, Roman Catholicism, Western liberalism, Eastern asceticism. See also the view of Milton Friedman in *Capitalism and Freedom* (Chicago: University of Chicago Press, 1962), p. 12: 'The liberal conceives of men as imperfect beings. He regards the problem of social organization to be as much a negative problem of preventing "bad" people from doing harm as of enabling "good" people to do good; and, of course, "bad" and "good" people may be the same people, depending on who is judging them.'

7 It is on just this point that Herbert Marcuse and other critics of Western liberal democracies have found liberalism so vulnerable. See R. Wolff et al., *A Critique of Pure Tolerance* (Boston: Beacon Press, 1965); Leo Strauss, *Natural Right and History* (Chicago: University of Chicago Press, 1953).

8 Above, note 6.

9 Hazel Barnes, *An Existentialist Ethics* (New York: Knopf, 1967). The influence of existentialism cannot be denied, especially in such fields as psychology and the practice of psychotherapy, which bear clearly on how individuals understand themselves as human beings.

10 As Christian Bay hints at above, note 5, David Hume has been the most influential proponent of the view that it is impossible to ground ethics on facts, for example about human nature. Generally, Hume's empiricist epistemology has had the utmost influence on the social sciences and philosophy itself, promoting the value-free stance concerning the proper approach to understanding human social life. I discuss some of these matters in my *Human Rights and Human Liberties* (Chicago: Nelson-Hall, 1975). See also my 'Epistemology and Moral Knowledge,' *The Review of Metaphysics*, September 1982, pp. 23–49.

One thing is crucial to remember about Hume's 'is–ought' gap argument – it is telling only against the idea that one can deduce moral claims from factual ones. Remember that he also argued that one cannot deduce scientific principles from facts. This is an argument against radical rationalism. As far as making inferences from claims about facts to claims about what we ought to do, on that score Hume is silent.

11 A while back, Robert Heilbroner, not usually considered a Stalinist in his sympathies for Marxism, put the matter as follows: 'A greater sense of economic liberty and a higher level of working morale may prevail in [socialism] than exists under capitalism. But it would be foolhardy – and, worse, deceptive – to expect that the transition itself could be made without recourse to massed command, or to deny that the transition will poise grave dangers for the abuse of that command or for its emplacement on a long-lasting basis' *Marxism* (New York: Norton, 1980), p. 159.

12 Renford Bambrough, *Moral Skepticism and Moral Knowledge* (Atlantic Highlands, NJ: Humanities Press, 1979), pp. 40ff.

13 For a good discussion of this point, see Douglas J. Den Uyl, 'Freedom and Virtue,' *Reason Papers*, Winter 1979: 1–12.

14 I leave aside a discussion of whether causing the death of innocent human beings correctly describes what individuals would be doing if their response to terrorism involved the death of innocent people. Arguably sometimes the deaths caused via retaliation could be construed as collateral, a necessary byproduct that the terrorists, as aggressors, caused. An excellent discussion of this and related issues may be found in a paper by Professor Eric Mack in Robert Poole, Jr (ed.), *Defending a Free Society* (Lexington, MA: Heath, 1984). See also, Tibor R. Machan, *The Passion for Liberty* (Lanham, MD: Rowman & Littlefield, 2003), Chapter 13: 'Military Defense of a Free Society.'

15 I wish to thank David Rapoport and J. Roger Lee for their comments on an earlier draft of this chapter.

Epilogue

The topic of objectivity is quite well traveled in philosophy and the sciences, including, especially, in the social sciences. In this work my main purpose has been to examine some of the recent attacks on the very possibility of objectivity, of our capacity to know the world and its various features as they are rather than as we may have been induced by various factors – such as our culture, upbringing, the constitution of our minds, our emotions, our economic situation, or some combination of these – to take it to be. Richard Rorty and Thomas Kuhn served as examples of those who are skeptics about the possibility of objectivity, while C. S. Peirce's basic ideas served to illustrate how objective understanding cannot be abandoned in a serious philosophical position.

I also showed how some of what I tried to defend in this book can have an impact on certain special areas of concern, such as jurisprudence, psychology, journalism, and related fields. In the last section, which is based on a 1987 paper I wrote on certain ethical implications of terrorism, I argue that terrorism cannot be adequately understood without admitting that some values are objective.

This book is by no means a comprehensive treatise on objectivity and only intimates, although rather tellingly, what such a treatise would very likely conclude about the subject. I have also omitted discussing the views of some important thinkers who have had a great deal to say about objectivity – Sir Karl Popper comes to mind among others. In my view Popper's fallibilism is an attempt to shirk a serious re-conceptualization of human knowledge not as something merely probable but as something that has no relationship to earlier – for example, Platonic and Cartesian – ways of viewing the topic. The best characterization of human knowledge along these lines would be to see knowledge as contextual, not timelessly fixed. This reflects best the way the concept of knowledge is deployed in successful day-to-day instances. There is, in other words, nothing even remotely probabilistic or fallibilistic about such knowledge. Of course, human beings are fallible, yes – they can believe they know when they in fact do not or fail to. But when they do know, that's *bona fide*, genuine knowledge that has no built-in problems as the term 'fallible' suggests.

A final note may be worth making about what 'objective' means when one speaks correctly about such knowledge or truth or reality. It is, as I make clear in the Introduction, nothing more mysterious than to believe that when we do know something, it is not a product of our feelings, wishes, prejudices, or the like but awareness of the way things are, independently of how our minds may at times distort them for us. Yes, our awareness is processed: we have a mind of a certain kind and when we understand by using it our understanding is achieved by reliance on its attributes, on what makes it the kind of mind it is. But there is no warrant here for believing that the process of understanding is a means for distorting what is to be understood. As Ayn Rand observed, 'The implicit, but unadmitted premise of modern philosophy is the notion that "true" knowledge must be acquired without any means of cognition, and that identity is the disqualifying element of consciousness.'[1] This is clearly evident in Richard Rorty's claim that we are unable to understand reality because we lack a 'skyhook – something which might lift us out of our beliefs.' As if our beliefs were not beliefs about the world but impositions on it that achieve the opposite of what we aim for by hiding the world from us.

So, what I have tried to do is to show the source of the skepticism about objectivity, namely, a misguided conception of what knowledge must be. Once this is given up, objectivity no longer necessarily escapes us but can, with concentration and patience, be the central attribute of how we know the world.

Note

1 Ayn Rand, *Introduction to Objectivist Epistemology*, expanded 2nd edn (New York: New American Library, 1990).

Bibliography

Agassi, J. (1963), 'Towards an Historiography of Science,' *History and Theory, Studies in the Philosophy of History*, Supplement 2.

Austin, J. L. (1961), 'Other Minds,' *Philosophical Papers*, Oxford: Oxford University Press.

Bastiat, F. (1963), *Selected Essays in Political Economy*, Princeton, NJ: Van Nostrand.

Beardsmore, R. W. (1969), *Moral Reasoning*, London: Schocken.

Boyle, J. M., Grisez, G., and Tollefsen, O. (1972), 'Determinism, Freedom and Self-referential Arguments', *The Review of Metaphysics*, 26: 3–37.

Caton, H. (1973), *The Origins of Subjectivity*, New Haven: Yale University Press.

Chein, I. (1972), *The Science of Behavior and the Image of Man*, New York: Basic Books.

Davidson, D. (1971), 'Agency,' in R. Binkley, R. Bronough, and A. Marras (eds), *Agent, Action, and Reason*, Toronto: University of Toronto Press.

Efron, R. (1967), 'Biology Without Consciousness and Its Consequences,' *Perspectives in Biology and Medicine*, 11: 9–36.

——— (1969), 'What is Perception?,' in *Boston Studies in the Philosophy of Science*, Vol. IV (ed. R. S. Cohen and M. W. Wartofsky), Dordrecht: D. Reidel, pp. 137–73.

——— (1970), 'The Measurement of Perceptual Duration,' *Studium Generale*, 23: 550–61.

Feyerabend, P. K. (1963), 'How to Be a Good Empiricist: A Plea for Tolerance in Matters Epistemological,' in *The Delaware Seminar in Philosophy of Science*, Vol. 2 (ed. B. Baumrin), New York: Wiley Interscience, pp. 3–39.

Fisk, M. (1974), *Nature and Necessity*, Bloomington: Indiana University Press.

Gerber, D. (1974), 'On Argumentum ad Hominem,' *The Personalist*, 55: 27.

Jordan, J. (1969), 'Determinism's Dilemma,' *The Review of Metaphysics*, 23: 48–66.

Koestler, A. (1964), *Act of Creation*, New York: Macmillan.

——— (1967), *The Ghost in the Machine*, New York: Macmillan.

Kordig, C. R. (1971), *The Justification of Scientific Change*, Dordrecht: D. Reidel.

Kuhn, T. S. (1962), *The Structure of Scientific Revolutions*, Chicago: University of Chicago Press.

——— (1970), *The Structure of Scientific Revolutions*, 2nd edn, Chicago: University of Chicago Press.

Lakatos, I. and Musgrave, A. (eds) (1970), *Criticism and the Growth of Knowledge*, Cambridge: Cambridge University Press.

Lehrer, K. (1975), 'Social Consensus and Rational Agniology', *Synthese*: 141–60.

Louch, A. R. (1966), *Explanation and Human Action*, Berkeley: University of California Press.

Machan, T. R. (1973), 'Wittgenstein and Metaethics,' *Journal of Thought*, 54: 252–7.

——— (1974a), *The Pseudo-Science of B. F. Skinner*, New Rochelle, NY: Arlington House.

——— (1974b), 'Kulm's Impossible Proof and the Moral Element in Scientific Explanations,' *Theory and Decision*, 4(3): 355–74.

——— (1975), *Human Rights and Human Liberties*, Chicago: Nelson-Hall.

——— and Zupan, M. L. (1975), 'Back to Being Reasonable', *Philosophy of Science*, 142(3): 307–10.

——— (1989), *Individuals and Their Rights*, La Salle, IL: Open Court Publishing.

——— (1998), *Classical Individualism*, London: Routledge.

——— (2000), *Initiative – Human Agency and Society*, Stanford, CA: Hoover Institution Press.

——— (2001), *Ayn Rand*, New York: Peter Lang.

——— (2003), *The Passion for Liberty*, Lanham, MD: Rowman & Littlefield.

Madden, E. H. (1969), 'The Third View of Causality', *The Review of Metaphysics*, 23: 67–84.

——— and Harre, R. (1973), 'In Defense of Natural Agents', *The Philosophical Quarterly*, 23: 117–32.

Marx, K. and Engels, F. (1968), *Selected Works*, New York: International Publ.

Merton, R. (1962), 'Priorities in Scientific Discovery: A Chapter in the Sociology of Science,' in B. Barber and W. Hirsch (eds), *The Sociology of Science*, New York: Free Press.

Miller, E. F. (1971), 'Hume's Contribution to Behavioral Science', *Journal of the History of the Behavioral Sciences*, 7: 154–68.

——— (1972), 'Positivism, Historicism, and Political Inquiry', *The American Political Science Review*, 66: 796–817.

Polanyi, M. (1969), *Knowing and Being*, Chicago: University of Chicago Press.

Rand, A. (1961), *The Virtue of Selfishness: A New Concept of Egoism*, New York: New American Library.

——— (1966), *Introduction to Objectivist Epistemology*, New York: The Objectivist Inc.

Rostand, J. (1960), *Error and Deception in Science*, New York: Harper and Row.

Schagrin, M. L. (1973), 'On Being Unreasonable', *The Philosophy of Science*, 40: 1–9.

Scheffler, I. (1967), *Science and Subjectivity*, Indianapolis: Bobbs-Merrill.

Schoeck, H. (1970), *Envy*, New York: Harcourt, Brace and Jovanovich.

——— and Wiggins, J. W. (1961), *Relativism and the Study of Man*, Princeton, NJ: Van Nostrand.

Skinner, B. F. (1953), *Science and Human Behavior*, New York: Macmillan.

——— (1976), *Particulars of My Life*, New York: Knopf.

Strauss, L. (1971), *Natural Right and History*, Chicago: University of Chicago Press.

Stroud, B. (1966), 'Wittgenstein and Logical Necessity,' in G. Pitcher (ed.), *Wittgenstein*, New York: Anchor Books, pp. 477–96.

Taylor, R. (1966), *Action and Purpose*, Englewood Cliffs, NJ: Prentice-Hall.

Toulmin, S. (1953), *The Philosophy of Science*, Cambridge: Cambridge University Press.

Trigg, R. (1973), *Reason and Commitment*, Cambridge: Cambridge University Press.

Truzzi, M. (1974), *Verstehen: Subjective Understanding in the Social Sciences*, Menlo Park, CA: Addison Wesley.

Wheatley, J. (1969), 'Reasons for Acting,' *Dialogue*, 7: 553–67.

Winch, P. (1967), *The Idea of a Social Science*, New York: Schocken Books.

Wittgenstein, L. (1953, 1973), *Philosophical Investigations*, New York: Macmillan.

——— (1969), *On Certainty*, New York: J. and J. Harper.

Yolton, J. (1973), 'Action, Metaphysics and Modality,' *American Philosophical Quarter*ly, 10: 71–85.

Index